Pitman Research Notes in Mathematics Series

Submission of proposals for consideration
Suggestions for publication, in the form of outlines and representative samples, are invited by the Editorial Board for assessment. Intending authors should approach one of the main editors or another member of the Editorial Board, citing the relevant AMS subject classifications. Alternatively, outlines may be sent directly to the publisher's offices. Refereeing is by members of the board and other mathematical authorities in the topic concerned, throughout the world.

Preparation of accepted manuscripts
On acceptance of a proposal, the publisher will supply full instructions for the preparation of manuscripts in a form suitable for direct photo-lithographic reproduction. Specially printed grid sheets are provided and a contribution is offered by the publisher towards the cost of typing. Word processor output, subject to the publisher's approval, is also acceptable.

Illustrations should be prepared by the authors, ready for direct reproduction without further improvement. The use of hand-drawn symbols should be avoided wherever possible, in order to maintain maximum clarity of the text.

The publisher will be pleased to give any guidance necessary during the preparation of a typescript, and will be happy to answer any queries.

Important note
In order to avoid later retyping, intending authors are strongly urged not to begin final preparation of a typescript before receiving the publisher's guidelines and special paper. In this way it is hoped to preserve the uniform appearance of the series.

Longman Scientific & Technical
Longman House
Burnt Mill
Harlow, Essex, UK
(tel (0279) 26721)

Titles in this series

1 Improperly posed boundary value problems
A Carasso and A P Stone
2 Lie algebras generated by finite dimensional ideals
I N Stewart
3 Bifurcation problems in nonlinear elasticity
R W Dickey
4 Partial differential equations in the complex domain
D L Colton
5 Quasilinear hyperbolic systems and waves
A Jeffrey
6 Solution of boundary value problems by the method of integral operators
D L Colton
7 Taylor expansions and catastrophes
T Poston and I N Stewart
8 Function theoretic methods in differential equations
R P Gilbert and R J Weinacht
9 Differential topology with a view to applications
D R J Chillingworth
10 Characteristic classes of foliations
H V Pittie
11 Stochastic integration and generalized martingales
A U Kussmaul
12 Zeta-functions: An introduction to algebraic geometry
A D Thomas
13 Explicit *a priori* inequalities with applications to boundary value problems
V G Sigillito
14 Nonlinear diffusion
W E Fitzgibbon III and H F Walker
15 Unsolved problems concerning lattice points
J Hammer
16 Edge-colourings of graphs
S Fiorini and R J Wilson
17 Nonlinear analysis and mechanics: Heriot-Watt Symposium Volume I
R J Knops
18 Actions of fine abelian groups
C Kosniowski
19 Closed graph theorems and webbed spaces
M De Wilde
20 Singular perturbation techniques applied to integro-differential equations
H Grabmüller
21 Retarded functional differential equations: A global point of view
S E A Mohammed
22 Multiparameter spectral theory in Hilbert space
B D Sleeman
24 Mathematical modelling techniques
R Aris
25 Singular points of smooth mappings
C G Gibson
26 Nonlinear evolution equations solvable by the spectral transform
F Calogero
27 Nonlinear analysis and mechanics: Heriot-Watt Symposium Volume II
R J Knops
28 Constructive functional analysis
D S Bridges
29 Elongational flows: Aspects of the behaviour of model elasticoviscous fluids
C J S Petrie
30 Nonlinear analysis and mechanics: Heriot-Watt Symposium Volume III
R J Knops
31 Fractional calculus and integral transforms of generalized functions
A C McBride
32 Complex manifold techniques in theoretical physics
D E Lerner and P D Sommers
33 Hilbert's third problem: scissors congruence
C-H Sah
34 Graph theory and combinatorics
R J Wilson
35 The Tricomi equation with applications to the theory of plane transonic flow
A R Manwell
36 Abstract differential equations
S D Zaidman
37 Advances in twistor theory
L P Hughston and R S Ward
38 Operator theory and functional analysis
I Erdelyi
39 Nonlinear analysis and mechanics: Heriot-Watt Symposium Volume IV
R J Knops
40 Singular systems of differential equations
S L Campbell
41 N-dimensional crystallography
R L E Schwarzenberger
42 Nonlinear partial differential equations in physical problems
D Graffi
43 Shifts and periodicity for right invertible operators
D Przeworska-Rolewicz
44 Rings with chain conditions
A W Chatters and C R Hajarnavis
45 Moduli, deformations and classifications of compact complex manifolds
D Sundararaman
46 Nonlinear problems of analysis in geometry and mechanics
M Atteia, D Bancel and I Gumowski
47 Algorithmic methods in optimal control
W A Gruver and E Sachs
48 Abstract Cauchy problems and functional differential equations
F Kappel and W Schappacher
49 Sequence spaces
W H Ruckle
50 Recent contributions to nonlinear partial differential equations
H Berestycki and H Brezis
51 Subnormal operators
J B Conway

52 Wave propagation in viscoelastic media
F Mainardi
53 Nonlinear partial differential equations and their applications: Collège de France Seminar. Volume I
H Brezis and J L Lions
54 Geometry of Coxeter groups
H Hiller
55 Cusps of Gauss mappings
T Banchoff, T Gaffney and C McCrory
56 An approach to algebraic K-theory
A J Berrick
57 Convex analysis and optimization
J-P Aubin and R B Vintner
58 Convex analysis with applications in the differentiation of convex functions
J R Giles
59 Weak and variational methods for moving boundary problems
C M Elliott and J R Ockendon
60 Nonlinear partial differential equations and their applications: Collège de France Seminar. Volume II
H Brezis and J L Lions
61 Singular systems of differential equations II
S L Campbell
62 Rates of convergence in the central limit theorem
Peter Hall
63 Solution of differential equations by means of one-parameter groups
J M Hill
64 Hankel operators on Hilbert space
S C Power
65 Schrödinger-type operators with continuous spectra
M S P Eastham and H Kalf
66 Recent applications of generalized inverses
S L Campbell
67 Riesz and Fredholm theory in Banach algebra
B A Barnes, G J Murphy, M R F Smyth and T T West
68 Evolution equations and their applications
F Kappel and W Schappacher
69 Generalized solutions of Hamilton-Jacobi equations
P L Lions
70 Nonlinear partial differential equations and their applications: Collège de France Seminar. Volume III
H Brezis and J L Lions
71 Spectral theory and wave operators for the Schrödinger equation
A M Berthier
72 Approximation of Hilbert space operators I
D A Herrero
73 Vector valued Nevanlinna Theory
H J W Ziegler
74 Instability, nonexistence and weighted energy methods in fluid dynamics and related theories
B Straughan
75 Local bifurcation and symmetry
A Vanderbauwhede
76 Clifford analysis
F Brackx, R Delanghe and F Sommen
77 Nonlinear equivalence, reduction of PDEs to ODEs and fast convergent numerical methods
E E Rosinger
78 Free boundary problems, theory and applications. Volume I
A Fasano and M Primicerio
79 Free boundary problems, theory and applications. Volume II
A Fasano and M Primicerio
80 Symplectic geometry
A Crumeyrolle and J Grifone
81 An algorithmic analysis of a communication model with retransmission of flawed messages
D M Lucantoni
82 Geometric games and their applications
W H Ruckle
83 Additive groups of rings
S Feigelstock
84 Nonlinear partial differential equations and their applications: Collège de France Seminar. Volume IV
H Brezis and J L Lions
85 Multiplicative functionals on topological algebras
T Husain
86 Hamilton-Jacobi equations in Hilbert spaces
V Barbu and G Da Prato
87 Harmonic maps with symmetry, harmonic morphisms and deformations of metrics
P Baird
88 Similarity solutions of nonlinear partial differential equations
L Dresner
89 Contributions to nonlinear partial differential equations
C Bardos, A Damlamian, J I Díaz and J Hernández
90 Banach and Hilbert spaces of vector-valued functions
J Burbea and P Masani
91 Control and observation of neutral systems
D Salamon
92 Banach bundles, Banach modules and automorphisms of C*-algebras
M J Dupré and R M Gillette
93 Nonlinear partial differential equations and their applications: Collège de France Seminar. Volume V
H Brezis and J L Lions
94 Computer algebra in applied mathematics: an introduction to MACSYMA
R H Rand
95 Advances in nonlinear waves. Volume I
L Debnath
96 FC-groups
M J Tomkinson
97 Topics in relaxation and ellipsoidal methods
M Akgül
98 Analogue of the group algebra for topological semigroups
H Dzinotyiweyi
99 Stochastic functional differential equations
S E A Mohammed

100 Optimal control of variational inequalities
V Barbu
101 Partial differential equations and dynamical systems
W E Fitzgibbon III
102 Approximation of Hilbert space operators. Volume II
C Apostol, L A Fialkow, D A Herrero and D Voiculescu
103 Nondiscrete induction and iterative processes
V Ptak and F-A Potra
104 Analytic functions – growth aspects
O P Juneja and G P Kapoor
105 Theory of Tikhonov regularization for Fredholm equations of the first kind
C W Groetsch
106 Nonlinear partial differential equations and free boundaries. Volume I
J I Díaz
107 Tight and taut immersions of manifolds
T E Cecil and P J Ryan
108 A layering method for viscous, incompressible L_p flows occupying R^n
A Douglis and E B Fabes
109 Nonlinear partial differential equations and their applications: Collège de France Seminar. Volume VI
H Brezis and J L Lions
110 Finite generalized quadrangles
S E Payne and J A Thas
111 Advances in nonlinear waves. Volume II
L Debnath
112 Topics in several complex variables
E Ramírez de Arellano and D Sundararaman
113 Differential equations, flow invariance and applications
N H Pavel
114 Geometrical combinatorics
F C Holroyd and R J Wilson
115 Generators of strongly continuous semigroups
J A van Casteren
116 Growth of algebras and Gelfand–Kirillov dimension
G R Krause and T H Lenagan
117 Theory of bases and cones
P K Kamthan and M Gupta
118 Linear groups and permutations
A R Camina and E A Whelan
119 General Wiener–Hopf factorization methods
F-O Speck
120 Free boundary problems: applications and theory, Volume III
A Bossavit, A Damlamian and M Fremond
121 Free boundary problems: applications and theory, Volume IV
A Bossavit, A Damlamian and M Fremond
122 Nonlinear partial differential equations and their applications: Collège de France Seminar. Volume VII
H Brezis and J L Lions
123 Geometric methods in operator algebras
H Araki and E G Effros
124 Infinite dimensional analysis–stochastic processes
S Albeverio
125 Ennio de Giorgi Colloquium
P Krée
126 Almost-periodic functions in abstract spaces
S Zaidman
127 Nonlinear variational problems
A Marino, L Modica, S Spagnolo and M Degiovanni
128 Second-order systems of partial differential equations in the plane
L K Hua, W Lin and C-Q Wu
129 Asymptotics of high-order ordinary differential equations
R B Paris and A D Wood
130 Stochastic differential equations
R Wu
131 Differential geometry
L A Cordero
132 Nonlinear differential equations
J K Hale and P Martinez-Amores
133 Approximation theory and applications
S P Singh
134 Near-rings and their links with groups
J D P Meldrum
135 Estimating eigenvalues with *a posteriori/a priori* inequalities
J R Kuttler and V G Sigillito
136 Regular semigroups as extensions
F J Pastijn and M Petrich
137 Representations of rank one Lie groups
D H Collingwood
138 Fractional calculus
G F Roach and A C McBride
139 Hamilton's principle in continuum mechanics
A Bedford
140 Numerical analysis
D F Griffiths and G A Watson
141 Semigroups, theory and applications. Volume I
H Brezis, M G Crandall and F Kappel
142 Distribution theorems of L-functions
D Joyner
143 Recent developments in structured continua
D De Kee and P Kaloni
144 Functional analysis and two-point differential operators
J Locker
145 Numerical methods for partial differential equations
S I Hariharan and T H Moulden
146 Completely bounded maps and dilations
V I Paulsen
147 Harmonic analysis on the Heisenberg nilpotent Lie group
W Schempp
148 Contributions to modern calculus of variations
L Cesari
149 Nonlinear parabolic equations: qualitative properties of solutions
L Boccardo and A Tesei
150 From local times to global geometry, control and physics
K D Elworthy

151 A stochastic maximum principle for optimal control of diffusions
U G Haussmann
152 Semigroups, theory and applications. Volume II
H Brezis, M G Crandall and F Kappel
153 A general theory of integration in function spaces
P Muldowney
154 Oakland Conference on partial differential equations and applied mathematics
L R Bragg and J W Dettman
155 Contributions to nonlinear partial differential equations. Volume II
J I Díaz and P L Lions
156 Semigroups of linear operators: an introduction
A C McBride
157 Ordinary and partial differential equations
B D Sleeman and R J Jarvis
158 Hyperbolic equations
F Colombini and M K V Murthy
159 Linear topologies on a ring: an overview
J S Golan
160 Dynamical systems and bifurcation theory
M I Camacho, M J Pacifico and F Takens
161 Branched coverings and algebraic functions
M Namba
162 Perturbation bounds for matrix eigenvalues
R Bhatia
163 Defect minimization in operator equations: theory and applications
R Reemtsen
164 Multidimensional Brownian excursions and potential theory
K Burdzy
165 Viscosity solutions and optimal control
R J Elliott
166 Nonlinear partial differential equations and their applications. Collège de France Seminar. Volume VIII
H Brezis and J L Lions
167 Theory and applications of inverse problems
H Haario
168 Energy stability and convection
G P Galdi and B Straughan
169 Additive groups of rings. Volume II
S Feigelstock
170 Numerical analysis 1987
D F Griffiths and G A Watson
171 Surveys of some recent results in operator theory. Volume I
J B Conway and B B Morrel
172 Amenable Banach algebras
J-P Pier
173 Pseudo-orbits of contact forms
A Bahri
174 Poisson algebras and Poisson manifolds
K H Bhaskara and K Viswanath
175 Maximum principles and eigenvalue problems in partial differential equations
P W Schaefer
176 Mathematical analysis of nonlinear, dynamic processes
K U Grusa
177 Cordes' two-parameter spectral representation theory
D F McGhee and R H Picard
178 Equivariant K-theory for proper actions
N C Phillips
179 Elliptic operators, topology and asymptotic methods
J Roe
180 Nonlinear evolution equations
J K Engelbrecht, V E Fridman and E N Pelinovski
181 Nonlinear partial differential equations and their applications. Collège de France Seminar. Volume IX
H Brezis and J L Lions
182 Critical points at infinity in some variational problems
A Bahri
183 Recent developments in hyperbolic equations
L Cattabriga, F Colombini, M K V Murthy and S Spagnolo
184 Optimization and identification of systems governed by evolution equations on Banach space
N U Ahmed
185 Free boundary problems: theory and applications. Volume I
K H Hoffmann and J Sprekels
186 Free boundary problems: theory and applications. Volume II
K H Hoffmann and J Sprekels
187 An introduction to intersection homology theory
F Kirwan
188 Derivatives, nuclei and dimensions on the frame of torsion theories
J S Golan and H Simmons
189 Theory of reproducing kernels and its applications
S Saitoh
190 Volterra integrodifferential equations in Banach spaces and applications
G Da Prato and M Iannelli
191 Nest algebras
K R Davidson
192 Surveys of some recent results in operator theory. Volume II
J B Conway and B B Morrel
193 Nonlinear variational problems. Volume II
A Marino and M K Murthy
194 Stochastic processes with multidimensional parameter
M E Dozzi
195 Prestressed bodies
D Iesan
196 Hilbert space approach to some classical transforms
R H Picard
197 Stochastic calculus in application
J R Norris
198 Radical theory
B J Gardner
199 The C* – algebras of a class of solvable Lie groups
X Wang

200 Stochastic analysis, path integration and dynamics
D Elworthy
201 Riemannian geometry and holonomy groups
S Salamon
202 Strong asymptotics for extremal errors and polynomials associated with Erdös type weights
D S Lubinsky
203 Optimal control of diffusion processes
V S Borkar
204 Rings, modules and radicals
B J Gardner
205 Numerical studies for nonlinear Schrödinger equations
B M Herbst and J A C Weideman
206 Distributions and analytic functions
R D Carmichael and D Mitrović
207 Semicontinuity, relaxation and integral representation in the calculus of variations
G Buttazzo
208 Recent advances in nonlinear elliptic and parabolic problems
P Bénilan, M Chipot, L Evans and M Pierre
209 Model completions, ring representations and the topology of the Pierce sheaf
A Carson
210 Retarded dynamical systems
G Stépán
211 Function spaces, differential operators and nonlinear analysis
L Päivärinta
212 Analytic function theory of one complex variable
Y Komatu, K Niino and C C Yang
213 Elements of stability of visco-elastic fluids
J Dunwoody
214 Jordan decompositions of generalised vector measures
K D Schmidt
215 A mathematical analysis of bending of plates with transverse shear deformation
C Constanda
216 Ordinary and partial differential equations Vol II
B D Sleeman and R J Jarvis
217 Hilbert modules over function algebras
R G Douglas and V I Paulsen
218 Graph colourings
R Wilson and R Nelson
219 Hardy-type inequalities
A Kufner and B Opic
220 Nonlinear partial differential equations and their applications. College de France Seminar Volume X
H Brezis and J L Lions
221 Workshop on dynamical systems
E Shiels and Z Coelho
222 Geometry and analysis in nonlinear dynamics
H W Broer and F Takens
223 Fluid dynamical aspects of combustion theory
M Onofri and A Tesei
224 Approximation of Hilbert space operators. Volume I 2nd edition.
D Herrero
225 Surveys of some recent results in operator theory. Volume III
J B Conway and B B Morrel
226 Local cohomology and localization
J L Bueso Montero, B Torrecillas Jover and A Verschoren
227 Sobolev spaces of holomorphic functions
F Beatrous and J Burbea
228 Numerical analysis. Volume III
D F Griffiths and G A Watson
229 Recent developments in structured continua Volume III
D De Kee and P Kaloni
230 Boolean methods in interpolation and approximation
F J Delvos and W Schempp

Boolean methods in interpolation and approximation

F–J Delvos and W Schempp

University of Siegen

Boolean methods in interpolation and approximation

Longman Scientific & Technical

Copublished in the United States with
John Wiley & Sons, Inc., New York

Longman Scientific & Technical,
Longman Group UK Limited,
Longman House, Burnt Mill, Harlow
Essex CM20 2JE, England
and Associated Companies throughout the world.

Copublished in the United States with
John Wiley & Sons, Inc., 605 Third Avenue, New York, NY 10158

First published 1989

AMS Subject Classification: (Main) 41A63, 41A05, 41A15
(Subsidiary) 41A80, 42B99, 63D32

ISSN 0269-3674

British Library Cataloguing in Publication Data
Delvos, F.J.
Boolean methods in interpolation and approximation.
1. Applied mathematics. Approximation
I. Title II. Schempp, W. (Walter), 1938–
511′.4

ISBN 0-582-06263-2

Library of Congress Cataloging-in-Publication Data
Delvos, F.J.
Boolean methods in interpolation and approximation/F–J Delvos and W. Schempp.
p. cm.—(Pitman research note in mathematics series: 230)
Includes bibliographical references.
ISBN 0-470-21583-6 (Wiley)
1. Interpolation. 2. Approximation theory. 3. Algebra, Boolean.
I. Schempp, W. (Walter), 1938–. II. Title. III. Series.
QA281.D45 1989
511′.42--dc20 89-13135
CIP

Printed and bound in Great Britain
by Biddles Ltd, Guildford and King's Lynn

Contents

Foreword

Multivariate interpolation and approximation have many important applications in the field of applied mathematics. As particular topics we mention the finite element method and computer aided geometric design. In the finite element method multivariate interpolation is used to construct sufficiently smooth piecewise polynomial function spaces with respect to a given triangulation of a domain where the variational problem connected with a boundary value problem has to be solved approximately. Computer aided geometric design is concerned with the construction of smooth surfaces which interpolate given curve networks. To solve this problem for the special case of a rectangular patch GORDON introduced the method of blending interpolation. He also started the investigation the Boolean algebraic background of blending interpolation.

It is the objective of this book to study the mathematical background of blending interpolation with respect to functional analysis and algebra. Moreover we want to treat some typical applications of Boolean methods to problems of bivariate interpolation and approximation.

In section 1 we start with commuting projectors in a linear space and show how to construct a smallest distributive lattice of commuting projectors containing the original ones. By adding complementary projectors to the original set of commuting projectors we construct a smallest Boolean algebra of commuting projectors containing the original set of projectors.

In section 2 we consider interpolation projectors which are bounded, idempotent, and linear mappings in a Banach algebra of complex or real valued continuous functions being defined on a compact metric space. Interpolation projectors are characterized by their kernels which are closed ideals in the Banach algebra of continuous functions. It is shown that the product of commuting interpolation projectors is again an interpolation projector. This fact enables the construction of distributive lattices of interpolation projectors. It is interesting to note that there are no Boolean algebras of interpolation projectors since a complementary interpolation projector cannot be an interpolation projector.

One basic tool of blending interpolation is the method of parametric extension. We treat this method in section 3 by considering tensor product operations in spaces of continuous functions which are defined on cartesian products of compact metric spaces. The Stone–Weierstraß theorem and its important special case, the Dieudonnée theorem, form the basic tools. Tensor products of bounded linear operators are most easily constructed and explicit formulas are obtained. In particular, a Fubini type theorem dealing with the commutativity of parametrically extended operators is established. This theorem is applied to parametric extensions of interpolation projectors and it turns out to be a basic method for constructing bivariate interpolation projectors from univariate ones. As simple instances the projectors of tensor product interpolation and blending interpolation are obtained.

In section 4 we investigate right invertible operators which are defined by projectors and injective operators. These operators are used to generate new projectors via an abstract Taylor formula. Boolean sums of parametrically extended projectors and tensor products of injective operators are used to construct tensor products of right invertible operators via the method of parametric extension. In this connection theorems of commutativity which are analogues of the theorems of Leibniz and Schwarz in classical analysis are established.

In section 5 the method of polynomial blending interpolation is deepened by constructing chains of bivariate polynomial interpolation projectors. As a typical example polynomial Lagrange interpolation on a regular triangular mesh is presented. This example is extended to a more sophisticated class of rectangular interpolation schemes which occur in the finite element method. Moreover the laws of the Boolean projector algebra are used to derive explicit remainder formulas for the different interpolation schemes.

The polynomial interpolation schemes treated in section 5 are usually applied as local approximation schemes. In section 6 we use the techniques of Boolean projector algebra to generate bivariate global approximation schemes based on periodic spline interpolation on uniform meshes. After comparing the approximation powers of spline blending interpolation and tensor product spline interpolation we introduce r–th order spline blending interpolation as an efficient alternative to tensor product spline interpolation.

In section 7 we introduce the univariate Korobov spaces of periodic functions via the method of right invertible operators presented in section 4.

The corresponding Taylor formula clarifies the interdependence of periodicity conditions and the asymptotic estimates of the Fourier coefficients of smooth functions. These properties are used to determine the approximation power of the univariate Fourier projector in Korobov spaces. We use the abstract approach of section 4 to construct the bivariate Korobov spaces. A Taylor formula for bivariate periodic functions enables the investigation of the relations between bivariate periodicity conditions and the asymptotic properties of the bivariate Fourier coefficients. We determine the approximation power of the Fourier tensor product projector and the Fourier blending projector in Korobov spaces. The discrete Fourier blending projector is shown to be more efficient than the Fourier product projector. The most natural bivariate Fourier approximation projector in Korobov spaces is the hyperbolic Fourier projector which is closely related to the Boolean construction of bivariate Korobov spaces. Using Boolean techniques we construct the somewhat simpler pseudo-hyperbolic Fourier projector and show that its approximation power is comparable with that of the hyperbolic Fourier projector.

As an application we construct cubature formulas for periodic functions via integration of interpolating splines. The Korobov spaces are used to establish practical error bounds for these formulas.

Siegen 1989

Franz-Jürgen Delvos
Walter Schempp

1 Lattices of projectors

1.1 Commuting projectors

Projectors form the fundamental tool in Boolean methods of interpolation and approximation. Here we list the basic definitions and algebraic properties such as Boolean sums of projectors and complementary pojectors.

Let X be a linear space over $\mathbb{R}$ or $\mathbb{C}$. A linear operator P on X is called a *projector* if and only if P satisfies the relation

$$P^2 = P .$$

The identity operator id_X on X is a projector which is denoted by I. The operator

$$P^c := I - P$$

is again a projector which is called the *remainder projector* or *complementary projector* of P. It is clear that the projectors P and P^c commute. We have the relations

$$P + P^c = I \ , \quad P^cP = 0 \ .$$

Here 0 denotes the *zero projector* on X. The *null space* of P is denoted by $\mathfrak{N}(P)$:

$$\mathfrak{N}(P) = \{ x \in X : Px = 0 \} \ .$$

The *range* of P is denoted by $\mathfrak{R}(P)$:

$$\mathfrak{R}(P) = \{ Px : x \in X \} \ .$$

Proposition 1. *If P is a projector, then*

(1) $$\mathfrak{R}(P^c) = \mathfrak{N}(P) \ .$$

Proof: If $y \in \mathfrak{R}(P^c)$, then $y = x - Px$ for some $x \in X$. Since $P^2 = P$ we get $Py = 0$. This implies $\mathfrak{R}(P^c) \subseteq \mathfrak{N}(P)$.

Assume $x \in \mathfrak{N}(P)$. Then we have $x = x - Px = P^c(x)$, i. e., we have $\mathfrak{N}(P) \subseteq \mathfrak{R}(P^c)$. This completes the proof of Proposition 1.

Let U and V be linear subspaces of X . Then

$$U + V = \{ x + y : x \in U , y \in V \}$$

is a linear subspace of X which is called the *sum* of U and V. U+V is called the *direct sum* of U and V if and only if the condition $U \cap V = \{0\}$ is satisfied.

Proposition 2. *If* P *is a projector on* X *, then* X *is the direct sum of* $\mathfrak{R}(P)$ *and* $\mathfrak{N}(P)$.

Proof: Recall that $\mathfrak{N}(P) = \mathfrak{R}(P^c)$. This implies $X = \mathfrak{R}(P) + \mathfrak{N}(P)$. If $x \in \mathfrak{R}(P) \cap \mathfrak{N}(P)$, then $x = Pz$ for some z and $Px = 0$ .Since $P^2 = P$ we obtain $0 = Px = P^2z = Pz = x$ which implies $\mathfrak{R}(P) \cap \mathfrak{N}(P) = \{0\}$ .This completes the proof of Proposition 2.

Let P and Q be *commuting* projectors :

$$PQ = QP \quad .$$

It is easily seen that the operator product of P and Q is also a projector which is called the *product projector* of P and Q .

Proposition 3. *If the projectors* P *and* Q *commute, then*

$$\mathfrak{R}(PQ) = \mathfrak{R}(P) \cap \mathfrak{R}(Q), \tag{2}$$

$$\mathfrak{N}(PQ) = \mathfrak{N}(P) + \mathfrak{N}(Q). \tag{3}$$

Proof: It is obvious from $PQ = QP$ that $\mathfrak{R}(PQ) \subseteq \mathfrak{R}(P) \cap \mathfrak{R}(Q)$ .If $z \in \mathfrak{R}()P \cap \mathfrak{R}(Q)$, then $z = Pz$, $z = Qz$ which implies $z = PQ(z)$ and $\mathfrak{R}(P) \cap \mathfrak{R}(Q) \subseteq \mathfrak{R}(PQ)$. This proves relation (2).

It follows from $PQ = QP$ that $\mathfrak{N}(P)+\mathfrak{N}(Q) \subseteq \mathfrak{N}(PQ)$. If $z \in \mathfrak{N}(PQ)$, then $Qz \in \mathfrak{N}(P)$. Moreover we have $z = Qz + Q^c z$ and $Q^c z \in \mathfrak{N}(Q)$. This implies $z \in \mathfrak{N}(P)+\mathcal{N}(Q)$ and therefore $\mathfrak{N}(PQ) \subseteq \mathfrak{N}(P)+\mathfrak{N}(Q)$ whence relation (3) follows. This completes the proof of Proposition 3.

Let P and Q be commuting projectors. It is easily seen that the remainder projectors P^c and Q^c alsc commute:

$$P^cQ^c = (I-P)(I-Q) = I-P-Q+PQ = I-P-Q+QP = (I-Q)(I-P) = Q^cP^c.$$

Thus, $(P^cQ^c)^c$ is also a projector which is given by

$$(P^cQ^c)^c = P + Q - PQ \quad . \tag{4}$$

This projector is called the *Boolean sum* of P and Q and is denoted by $P \oplus Q$:

$$P \oplus Q = P + Q - PQ \; .$$

Proposition 4. *If the projectors* P *and* Q *commute, then*

$$\mathfrak{R}(P \oplus Q) = \mathfrak{R}(P) + \mathfrak{R}(Q), \tag{5}$$

$$\mathfrak{N}(P \oplus Q) = \mathfrak{N}(P) \cap \mathfrak{N}(Q). \tag{6}$$

Proof: Taking into account the relations (4,), (1), (2), and (3) we obtain

$$\mathfrak{R}(P \oplus Q) = \mathfrak{R}(I - P^cQ^c) = \mathfrak{N}(P^cQ^c) = \mathfrak{N}(P^c) + \mathfrak{N}(Q^c) = \mathfrak{R}(P) + \mathfrak{R}(Q)$$

whence (5) follows.

To establish (6) we proceed in a similar way:

$$\mathfrak{N}(P\oplus Q) = \mathfrak{N}(I-P^cQ^c) = \mathfrak{R}(P^cQ^c) = \mathfrak{R}(P^c)\cap\mathfrak{R}(Q^c) = \mathfrak{N}(P)\cap\mathfrak{N}(Q)$$

whence (6) follows. This completes the proof of Proposition 4.

Let P and Q be commuting projectors. P and Q are called *orthogonal* if and only if $PQ = 0$. In this case we have

$$P \oplus Q = P + Q .$$

1.2 Construction of projector lattices

The algebraic operations of products and Booleans sums yield new projectors. The variety of these projectors is a source for interpolation and approximation projectors in particular in the multivariate setting. We present a lattice theoretical description of this procedure.

We assume that

$$\mathcal{L} = \{ A_j : j \in J \}$$

is a set of commuting projectors on a linear space:

$$A_j^2 = A_j \quad , \quad A_j A_k = A_k A_j \qquad (j,k \in J).$$

We enlarge $\mathcal{L}$ to the set $\mathcal{L}'$ by adding all possible products of projectors from $\mathcal{L}$:

$$\mathcal{L}' = \{ A_j A_k : j,k \in J \} \quad .$$

Since $A_j A_j = A_j$ we have

(1) $$\mathcal{L} \subseteq \mathcal{L}' \quad .$$

Proposition 1. $\mathcal{L}'$ *is a set of commuting projectors which contains the generating set* $\mathcal{L}$.

Proof: Since A_j and A_k are commuting projectors the product

$$A_{jk} = A_j A_k$$

is also a projector. We have

$$A_{jk} A_{lm} = A_j A_k A_l A_m = A_j A_l A_k A_m = A_l A_j A_m A_k = A_l A_m A_j A_k = A_{lm} A_{jk} \ .$$

This completes the proof of Proposition 1.

Next we enlarge $\mathscr{L}'$ to the set $\mathscr{L}''$ by adding all possible Boolean sums of projectors from $\mathscr{L}'$:

$$\mathscr{L}'' = \{ A_{jk} \oplus A_{lm} : j,k,l,m \in J \} .$$

Since $A_{jk} \oplus A_{jk} = A_{jk}$ we have

$$\mathscr{L}' \subseteq \mathscr{L}'' . \tag{2}$$

Proposition 2. *$\mathscr{L}''$ is a set of commuting projectors which contains the set $\mathscr{L}'$.*

Proof: Since $A_{jk}, A_{lm} \in \mathscr{L}'$, they are commuting projectors and therefore

$$A_{jklm} := A_{jk} \oplus A_{lm}$$

is also a projector. We will need the following

Lemma 1. *Let* A, B, C, *and* D *be commuting projectors. Then we have*

$$(A \oplus B)(C \oplus D) = (C \oplus D)(A \oplus B) . \tag{3}$$

Proof: We carry out the easy computations:

$$\begin{aligned} &(A \oplus B)(C \oplus D) \\ &= (A+B-AB)(C+D-CD) = AC+AD-ACD + BC+BD-BCD - ABC-ABD+ABCD \\ &= C(A \oplus B) + D(A \oplus B) - CD(A \oplus B) = (C \oplus D)(A \oplus B) . \end{aligned}$$

Now we can conclude

$$A_{jklm}\,A_{stuv} \;=\; (A_{jk}\oplus A_{lm})(A_{st}\oplus A_{uv}) \;=\; (A_{st}\oplus A_{uv})(A_{jk}\oplus A_{lm}) \;=\; A_{stuv}\,A_{jklm}$$

for $j,k,l,m,s,t,u,v \in J$. This completes the proof of Proposition 2.

Remark 1. It is easily seen from the definitions of $\mathscr{L}'$ and $\mathscr{L}''$ that if $\mathscr{L}$ is finite then $\mathscr{L}'$ and $\mathscr{L}''$ are also finite.

We apply the previous constructions to the set of commuting projectors

$$\mathscr{L}_1 = \mathscr{L}''$$

and iterate this process. We obtain an increasing sequence of sets of commuting projectors:

$$\mathscr{L}_{n+1} := \mathscr{L}''_n \qquad (n \geq 1) .$$

We have

(4) $$\mathscr{L}_n \subseteq \mathscr{L}_{n+1} \quad .$$

Proposition 3. *Let*

$$\mathfrak{L} = \bigcup_{n=1}^{\infty} \mathscr{L}_n \quad .$$

Then $\mathfrak{L}$ *is a set of commuting projectors with the properties that*

(5) $$PQ \in \mathfrak{L},$$

(6) $$P\oplus Q \in \mathfrak{L}$$

for all $P, Q \in \mathfrak{L}$.

Proof: If $A, B \in \mathfrak{L}$, then there is a positive integer m such that

$$AB \in \mathcal{L}_m \quad .$$

Taking into account Propositions 1 and 2 we obtain $AB = BA$ .Moreover we have $AB \in \mathcal{L}'_m$,$A \oplus B \in \mathcal{L}''_m$ which implies

$$AB \ , \ A \oplus B \in \mathfrak{L} \quad .$$

This completes the proof of Proposition 3.

It follows from the construction of $\mathfrak{L}$ that $\mathfrak{L}$ is the smallest set of commuting projectors containing $\mathcal{L}$ which is closed with respect to the operations of product and Boolean sum of projectors.

Proposition 4. *If* $A, B, C \in \mathfrak{L}$, *then*

(7) $$AB = BA \quad ,$$

(8) $$A \oplus B = B \oplus A \quad ,$$

(9) $$A(BC) = (AB)C \quad ,$$

(10) $$(A \oplus B) \oplus C = A \oplus (B \oplus C) \quad .$$

Proof: We establish (10) :

$$(A \oplus B) \oplus C$$

$$= A \oplus B + C - (A \oplus B)C = A+B-AB + C- AC-BC+ABC$$

$$= A + (B+C-BC) - A(B+C-BC) = A \oplus (B \oplus C) \quad .$$

This completes the proof of Proposition 4.

Proposition 5. *$\mathfrak{L}$ is a partially ordered set with respect to the order relation*

$$A \leq B \Leftrightarrow AB = A$$

for $A, B \in \mathfrak{L}$.

Proof: . It easily seen that the following relations hold:

$$(11) \qquad A \leq B \;,\; B \leq A \;\Rightarrow\; A = B \;,$$

$$(12) \qquad A \leq A \;.$$

We will verify

$$(13) \qquad A \leq B \;,\; B \leq C \;\Rightarrow\; A \leq C.$$

We have

$$AB = BA = A \;,\; BC = CB = B$$

which implies

$$AC = ABC = AB = A \;,$$

i. e., $A \leq C$. This completes the proof of Proposition 5.

Proposition 6. *$\mathfrak{L}$ is a lattice, i. e., the relations*

$$(14) \qquad \inf\{A,B\} = AB \;,$$

$$(15) \qquad \sup\{A,B\} = A \oplus B$$

are valid for every $A, B \in \mathfrak{L}$.

Proof: It is obvious that

$$AB \leq A \;,\; AB \leq B \;.$$

Assume that $C \in \mathfrak{L}$ satisfies

$$C \leq A \quad , \quad C \leq B \; .$$

Then we have

$$CA = C \quad , \quad AB = C$$

which implies $CAB = C$ whence relation (14) follows.

It is easily seen that

$$A \leq A \oplus B \quad , \quad B \leq A \oplus B \quad .$$

Assume that $D \in \mathfrak{L}$ satisfies

$$A \leq D \quad , \quad B \leq D \quad ,$$

i.e., we have

$$AD = A \quad , \quad BD = B \quad .$$

Then we obtain

$$(A \oplus B)D = AD + BD - ABD = A + B - A \quad ,$$

i. e., we have $A \oplus B \leq D$. This implies (15) and completes the proof of Proposition 6.

Proposition 7. $\mathfrak{L}$ *is a distributive lattice, i. e., the relations*

$$A(B \oplus C) = (AB) \oplus (AC), \tag{16}$$

$$A \oplus (BC) = (A \oplus B)(A \oplus C) \tag{17}$$

hold for every $A, B, C \in \mathfrak{L}$.

Proof: We have

$$A(B\oplus C) = AB + AC - ABC = AB + AC - ABAC = (AB)\oplus(AC)$$

whence (16) follows.
Next we obtain

$$A\oplus(BC)$$

$$= A + BC - ABC = A + AC - AC + AB + BC - ABC - AB - ABC + ABC$$

$$= (A + B - AB)(A + C - AC) = (A\oplus B)(A\oplus C)$$

whence (17) follows. This completes the proof of Proposition 7.

Remark 2. It follows from the definition of $\mathfrak{L}$ an Proposition 7 that the distributive lattice $\mathfrak{L}$ is finite if the *generating set* $\mathscr{L}$ of commuting projectors is finite.

1.3 Chains of projectors

Many projectors in the theory of interpolation and approximation form chains with respect to the partial ordering described in Section 1.2. We will show how to generate a new chain of projectors from to given chains of projectors. This procedure will be used in the construction of multivariate projectors of interpolation and approximation.

Let $\mathfrak{L}$ be a distributive lattice of commuting projectors on a linear space. We assume that the subsets

$$\{ A_j : j \in \mathbb{N} \} \quad , \quad \{ B_k : k \in \mathbb{N} \}$$

of $\mathfrak{L}$ form chains, i. e., we have

$$A_j \leq A_{j+1} \quad (j \in \mathbb{N}) \quad , \quad B_k \leq B_{k+1} \quad (k \in \mathbb{N}) \quad .$$

Using these chains we define a new sequence of projectors in $\mathfrak{L}$ by

$$C_1 = A_1B_1 \quad ,$$

$$C_2 = A_1B_2 \oplus A_2B_1 \quad ,$$

$$\vdots$$

$$C_r = A_1B_r \oplus A_2B_{r-1} \oplus \cdots \oplus A_rB_1$$

where $r \in \mathbb{N}$.

Proposition 1. *The set of projectors* $\{ C_r : r \in \mathbb{N} \}$ *form a chain in* $\mathfrak{L}$.

Proof: It follows from Proposition 6 of Section 1.2 that

$$C_r = \sup\{ A_mB_{r+1-m} : 1 \leq m \leq r \} \quad .$$

Since

$$B_j \leq B_{j+1} \quad (j \in \mathbb{N})$$

we have

$$A_jB_{r+1-j} \leq A_jB_{r+2-j} \quad (1 \leq j \leq r)$$

which implies

$$\begin{aligned} C_{r+1} &= \sup\{ A_mB_{r-2-m} : 1 \leq m \leq r+1 \} \\ &\geq \sup\{ A_mB_{r+2-m} : 1 \leq m \leq r \} \\ &\geq \sup\{ A_mB_{r+1-m} : 1 \leq m \leq r \} \\ &= C_r \quad . \end{aligned}$$

This completes the proof of Proposition 1.

Proposition 2. *The chain projector* C_r *possesses the explicit sum representation*

(1) $$C_r = \sum_{m=1}^{r} A_mB_{r+1-m} - \sum_{m=1}^{r-1} A_mB_{r-m} \quad .$$

Proof: We start with $r = 2$. Since

$$A_1A_2 = A_1 \quad , \quad B_1B_2 = B_1$$

we obtain

$$\begin{aligned} C_2 = A_1B_2 \oplus A_2B_1 &= A_1B_2 + A_2B_1 - A_1B_2A_2B_1 = A_1B_2 + A_2B_1 - A_1A_2B_2B_1 \\ &= A_1B_2 + A_2B_1 - A_1B_1 \quad . \end{aligned}$$

We proceed by induction :

$$C_{r+1}$$

$$= A_1B_{r+1} \oplus \cdots \oplus A_rB_2 \oplus A_{r+1}B_1$$

$$= A_1B_{r+1} \oplus \cdots \oplus A_rB_2 + A_{r+1}B_1 - (A_1B_{r+1} \oplus \cdots \oplus A_rB_2)A_{r+1}B_1$$

$$= A_1B_{r+1} + \cdots + A_{r+1}B_1 - (A_1B_r + \cdots + A_{r-1}B_2)$$

$$- (A_1A_{r+1}B_{r+1}B_1) \oplus \cdots \oplus (A_rA_{r+1}B_2B_1)$$

$$= A_1B_{r+1} + \cdots + A_{r+1}B_1 - (A_1B_r + \cdots + A_{r-1}B_2) - A_1B_1 \oplus \cdots \oplus A_rB_1$$

$$= A_1B_{r+1} + \cdots + A_{r+1}B_1 - (A_1B_r + \cdots + A_rB_1)$$

wchich completes the proof of Proposition 2.

It is easily seen that the projectors

$$A_1(B_r - B_{r-1}), \cdots , A_{r-1}(B_2 - B_1) , A_rB_1$$

are pairwise orthogonal. It follows from relation (1) that

$$A_1B_r \oplus \cdots \oplus A_{r-1}B_2 \oplus A_rB_1 = A_1(B_r - B_{r-1}) \oplus \cdots \oplus A_{r-1}(B_2 - B_1) \oplus A_rB_1 \tag{2}$$

holds. Relation (2) is called the *orthogonal decomposition* of the chain projector C_r .

1.4 Construction of Boolean algebras of projectors

Complementary projectors occur in remainder formulas of interpolation and approximation processes. It is shown that the distributive lattices whose generating sets are closed with respect to the operation of complementation are Boolean algebras of commuting projectors. Owing to their importance complementary chain projectors are studied in greater detail.

Let

$$\mathcal{L} = \{ A_j : j \in J \}$$

be a set of commuting projectors on a linear space X. We introduce the set

$$\mathcal{L}^c = \{ A_j^c : j \in J \} .$$

Then

$$\mathcal{K} = \mathcal{L} \cup \mathcal{L}^c$$

is also a set of commuting projectors. It possesses the property

$$P \in \mathcal{K} \Rightarrow P^c \in \mathcal{K} \tag{1}$$

Let $\mathfrak{K}$ be the distributive lattice of commuting projectors generated from $\mathcal{K} = \mathcal{L} \cup \mathcal{L}^c$. Recall that

$$\mathfrak{K} = \bigcup_{r=1}^{\infty} \mathcal{K}_r$$

with

$$\mathcal{K}_1 = \mathcal{K}'' , \quad \mathcal{K}_{r+1} = \mathcal{K}_r'' \quad (r \geq 1) .$$

Lemma 1. *The relation*

$$P \in \mathcal{K}_r \Rightarrow P^c \in \mathcal{K}_r' \tag{2}$$

holds for every positive integer r .

Proof: We consider the case $r = 1$. If $P \in \mathcal{K}_1$, then there exist projectors $A, B, C, D \in \mathcal{K}$ such that

$$P = AB \oplus CD .$$

Taking into account relation (4) of Section 1.1 we get

$$P^c = (AB)^c(CD)^c = (A^c \oplus B^c)(C^c \oplus D^c) .$$

It follows from relation (1) that

$$A^c \oplus B^c , C^c \oplus D^c \in \mathcal{K}_1 = \mathcal{K}''$$

which implies

$$P^c \in \mathcal{K}'_1 .$$

We proceed by induction. If $P \in \mathcal{K}_{r+1}$, then P possesses the representation

$$P = AB \oplus CD$$

with $A, B, C, D \in \mathcal{K}_r$. Then we have

$$P^c = (A^c \oplus B^c)(C^c \oplus D^c)$$

with

$$A^c, B^c, C^c, D^c \in \mathcal{K}'_r .$$

Now we can conclude

$$A^c \oplus B^c, C^c \oplus D^c \in \mathcal{K}''_r = \mathcal{K}_{r+1}$$

which implies

$$P^c \in \mathcal{K}'_{r+1} .$$

This completes the proof of Lemma 1 .

Proposition 1. *The distributive lattice* $\mathfrak{K}$ *of commuting projectors generated from* $\mathcal{K} = \mathcal{L} \cup \mathcal{L}^c$ *is complemented, i. e., we have*

$$P \in \mathfrak{K} \Rightarrow P^c \in \mathfrak{K} . \tag{3}$$

Proof: Relation (3) is a consequence of Lemma 1. Assume $P \in \mathfrak{K}$. Then there is a positive integer such that $P \in \mathcal{K}_r$. By Lemma 1 we obtain $P^c \in \mathcal{K}'_r$. Since $\mathcal{K}'_r \subseteq \mathcal{K}_{r+1} \subseteq \mathfrak{K}$, the proof of Proposition 1 is complete.

Remark 1. Since $\mathcal{K} = \mathcal{L} \cup \mathcal{L}^c$, $\mathfrak{K}$ is also called the *Boolean algebra of commuting projectors generated from* $\mathcal{L}$. $\mathfrak{K}$ is finite if $\mathcal{L}$ is finite.

Assume that

$$\{ A_j : j \in \mathbb{N} \} \quad , \quad \{ B_k : k \in \mathbb{N} \}$$

are chains in $\mathfrak{K}$. It follows from Proposition 1 of Section 1.3 that

$$\{ A_1 B_r \oplus \cdots \oplus A_r B_1 : r \in \mathbb{N} \}$$

is a chain in the Boolean algebra $\mathfrak{K}$. We will determine the remainder projectors

$$I - A_1 B_r \oplus \cdots \oplus A_r B_1 \qquad (r \in \mathbb{N})$$

which are also elements of $\mathfrak{K}$.

Proposition 2. *The relation*

$$(A_1 B_r \oplus \cdots \oplus A_r B_1)^c = A_r^c + A_{r-1}^c B_1^c + \cdots + A_1^c B_{r-1}^c + B_r^c - (A_r^c B_1^c + \cdots + A_1^c B_r^c). \tag{4}$$

is true for all $r \in \mathbb{N}$.

Proof: It follows from Proposition 2 of Section 1.3 that

$$A_1B_r \oplus \cdots \oplus A_rB_1 = A_1B_r + \cdots + A_rB_1 - (A_1B_{r-1} + \cdots + A_{r-1}B_1) .$$

Then we can conclude

$$\begin{aligned}
& I - A_1B_r \oplus \cdots \oplus A_rB_1 \\
= \quad & A_1B_{r-1} + \cdots + A_{r-1}B_1 - (A_1B_r + \cdots + A_rB_1) + I \\
= \quad & (I-A_1^c)(I-B_{r-1}^c) + \cdots + (I-A_{r-1}^c)(I-B_1^c) \\
& - (I-A_1^c)(I-B_r^c) + \cdots + (I-A_r^c)(I-B_1^c) + I \\
= \quad & A_1^c + \cdots + A_r^c + B_1^c + \cdots + B_r^c \\
& - (A_1^cB_r^c + \cdots + A_r^cB_1^c) + (A_1^cB_{r-1}^c + \cdots + A_{r-1}^cB_1^c) \\
& - (A_1^c + \cdots + A_{r-1}^c) - (B_1^c + \cdots + B_{r-1}^c) \\
= \quad & A_r^c + A_1^cB_{r-1}^c + \cdots + A_{r-1}^cB_1^c + B_r^c \\
& - (A_1^cB_{r-1}^c + \cdots + A_{r-1}^cB_1^c) .
\end{aligned}$$

This completes the proof of Proposition 2.

Notes, Section 1

The basic paper on projector lattices is GORDON [28]. Our constructive approach is based on GORDON - CHENEY [30] (see also BASZENSKI-DELVOS [4]). A different method for constructing Boolean algebras of projectors is described in DELVOS [12].

2 Interpolation projectors

2.1 Closed ideals

The null space of the polynomial Lagrange interpolation projector acting on an algebra of continuous functions being defined on a compact interval is closed ideal of this algebra. This connection between ideals and interpolation will be investigated in the general framework of Banach algebras of continuous functions being defined on a compact metric space. In this section we collect some properties of closed ideals in Banach algebras of continuous functions. The concept of zero set of a closed ideal turns out to be of basic importance.

Let E be a compact metric space with *distance* d . We denote by $\mathcal{C}(E)$ the linear space of continuous functions on E having real (or complex) function values. $\mathcal{C}(E)$ is a Banach algebra with respect to the uniform norm:

$$\|f\|_{\infty} = \sup \{ |f(t)| : t \in E \} \quad (f \in \mathcal{C}(E)).$$

A linear subspace U of $\mathcal{C}(E)$ is called an *ideal* if and only if the condition

$$f \in \mathcal{C}(E) \ , \ g \in U \ \Rightarrow \ fg \in U$$

holds. Instances of ideals are described in the following

Proposition 1. *Let* T *be a subset of* E *and let*

$$\mathcal{C}(E\|T) = \{ f \in \mathcal{C}(E) : f(t) = 0 \text{ for every } t \in T \} .$$

Then $\mathcal{C}(E\|T)$ *is an ideal in* $\mathcal{C}(E)$ *which is also closed.*

Proof: Assume that $f \in \mathcal{C}(E)$ and $g \in \mathcal{C}(E\|T)$. Then we have

$$f(t)g(t) = 0 \quad (t \in T)$$

which implies $fg \in \mathcal{C}(E\|T)$. It is obvious that $\mathcal{C}(E\|T)$ is a linear subspace of $\mathcal{C}(E)$. Therefore $\mathcal{C}(E\|T)$ is an ideal in $\mathcal{C}(E)$.

Let f_n , $n \in \mathbb{N}$, be a sequence in $\mathcal{C}(E\|T)$ which converges in the uniform norm to $f \in \mathcal{C}(E)$. Since uniform convergence implies pointwise convergence we can conclude

$$f(t) = 0 \quad (t \in T) .$$

This completes the proof of Proposition 1.

Let $\overline{T}$ denote the closure of T . It is easily seen that the relation

$$\mathcal{C}(E\|T) = \mathcal{C}(E\|\overline{T})$$

holds. The trivial ideals $\mathcal{C}(E)$ and $\{0\}$ are characterized by the sets E and $\emptyset$, respectively:

$$\mathcal{C}(E\|E) = \{0\} , \quad \mathcal{C}(E\|\emptyset) = \mathcal{C}(E) .$$

Proposition 2. *Let* S *and* T *be closed subsets of* E. *Then the relation*

$$\mathcal{C}(E\|S) \cap \mathcal{C}(E\|T) = \mathcal{C}(E\|S\cup T) \tag{1}$$

is valid.

Proof: It is easily seen that $\mathcal{C}(E\|S) \cap \mathcal{C}(E\|T)$ is a closed ideal with the property that

$$\mathcal{C}(E\|S)\cap\mathcal{C}(E\|T) \subseteq \mathcal{C}(E\|S\cup T) .$$

On the other hand, we have

$$\mathscr{C}(E\|S\cup T) \subseteq \mathscr{C}(E\|S)\cap\mathscr{C}(E\|T)$$

which completes the proof of Proposition 2.

Let U be a closed ideal in $\mathscr{C}(E)$. Then the set

$$(2) \qquad \mathcal{Z}(U) = \{ t \in E : g(t) = 0 \text{ for all } g \in U \}$$

is called the *zero set* of U .

Proposition 3. *If* U *is a closed ideal in* $\mathscr{C}(E)$, *then the relation*

$$(3) \qquad U = \mathscr{C}(E\|\mathcal{Z}(U))$$

is valid.

Proof: Note first that

$$U \subseteq \mathscr{C}(E\|\mathcal{Z}(U)) .$$

Moreover, the zero set $\mathcal{Z}(U)$ is a closed subset of E . We will need the following

Lemma 1. *Let* T *be a closed subset of* E *with the property that*

$$T \cap \mathcal{Z}(U) = \emptyset .$$

Then there is a function $h \in U$ *satisfying*

$$h(t) = 1 \quad (t \in T) , \quad 0 \leq h(x) \leq 1 \quad (x \in E) .$$

Proof: We follow NACHBIN [39]. For every $s \in T$ there is a function $h_s \in$

U with the property that

$$h_s(s) = 1 \ , \ h_s(x) \geq 0 \quad (x \in E) .$$

The set

$$V_s = \{ x \in E : h_s(x) > 0 \}$$

is open. Since T is compact, there exist $s_1, \cdots, s_n \in T$ with the property that

$$T \subseteq \bigcup_{j=1}^{n} V_{s_j} .$$

The function

$$H = h_{s_1} + \cdots + h_{s_n}$$

is an element of the ideal which is positive on T . There exist a positive constant c with the property that

$$\inf\{ H(t) : t \in T \} = c > 0 .$$

By the Tietze - Urysohn extension theorem there is a function $f \in \mathcal{C}(E)$ satisfying

$$k(t) = 1/H(t) \quad (t \in T)$$

and

$$\inf\{ 1/H(t) : t \in T \} \leq k(x) \leq \sup\{ 1/H(t) : t \in T \} \quad (x \in E).$$

Then the function $K \in \mathcal{C}(E)$ being defined by

$$K(x) = \inf\{ k(x), 1/H(x)\}$$

satisfies

$$K(t) = 1/H(t) \quad (t \in T) \ , \quad 0 \leq K(x) \leq 1/H(x) \quad (x \in E) .$$

Then $h := KH$ satisfies the relations

$$h \in U \quad , \quad 0 \le h(x) \le 1 \quad (x \in E) \quad , \quad h(t) = 1 \quad (t \in T) \quad .$$

This completes the proof of Lemma 1.

Let $\varepsilon > 0$ and $f \in \mathscr{C}(E \| \mathcal{Z}(U))$. Then

$$T_\varepsilon \quad := \quad \{ x \in E : |f(x)| \ge \varepsilon \}$$

is a closed set satisfying

$$T_\varepsilon \cap \mathcal{Z}(U) \quad = \quad \emptyset \quad .$$

Choose $h_\varepsilon \in U$ corresponding to T_ε as described in Lemma 1. Then we have

$$g_\varepsilon \quad = \quad fh_\varepsilon \quad \in \quad U$$

and

$$|f(x) - g_\varepsilon(x)| \quad = \quad |f(x)|(1 - h_\varepsilon(x)) \quad \le \quad \varepsilon \quad (x \in E) \quad ,$$

whence $\|f - g_\varepsilon\| \le \varepsilon$. Thus we have shown $f \in \bar{U} = U$, i. e., $\mathscr{C}(E \| \mathcal{Z}(U)) \subseteq U$. This completes the proof of Proposition 3.

As an application we prove the following

Proposition 4. *Let* S *and* T *be closed subsets of* E *such that the ideal* $\mathscr{C}(E\|S) + \mathscr{C}(E\|T)$ *is closed. Then*

$$\mathscr{C}(E\|S) + \mathscr{C}(E\|T) \quad = \quad \mathscr{C}(E\|S \cap T). \tag{4}$$

Proof: Let X be the zero set of $\mathscr{C}(E\|S) + \mathscr{C}(E\|T)$:

$$X \quad = \quad \mathcal{Z}(\mathscr{C}(E\|S) + \mathscr{C}(E\|T)) \quad .$$

It follows from Proposition 3 that

$$\mathcal{C}(E\|S) + \mathcal{C}(E\|T) \quad = \quad \mathcal{C}(E\|X) \quad .$$

This implies

$$S \cap T \subseteq X \quad .$$

We consider the functions $f \in \mathcal{C}(E\|S)$ and $g \in \mathcal{C}(E\|T)$ which are distance functions for the closed sets S and T , respectively:

$$f(x) \quad = \quad \inf\{\ d(x,s) : s \in S\ \} \quad , \quad g(x) \quad = \quad \inf\{\ d(x,t) : s \in T\ \}.$$

Then we have

$$f + g \quad \in \quad \mathcal{C}(E\|S) + \mathcal{C}(E\|T) \quad = \quad \mathcal{C}(E\|X) \ ,$$

$$f(x) + g(x) \ = \ 0 \quad \Leftrightarrow \quad x \in S \cap T$$

which implies

$$X \ = \ S \cap T \quad .$$

This completes the proof of Proposition 4.

2.2 Precision sets

The set of interpolation points (the *precision set*) of the polynomial Lagrange interpolation projector coincides with the zero set of its null space. An abstract interpolation projector is a bounded projector acting on a Banach algebra of continuous functions such that its null space is an ideal and the constant functions are left invariant. For these projectors we will show that zero set of the null space and the precision set are identical.

Let P be a bounded projector $\mathcal{C}(E)$. The *precision set* of P is defined by

$$\mathcal{P}(P) \quad = \quad \{ x \in E : \; P(f)(x) = f(x) \text{ for all } f \in \mathcal{C}(E) \} ,$$

i. e., $\mathcal{P}(P)$ is the *set of interpolation points* of the projector P . We have

$$\mathcal{P}(I) \; = \; E \quad , \quad \mathcal{P}(0) \; = \; \emptyset \; .$$

It is obvious that the precision set $\mathcal{P}(P)$ is a closed subset of the compact metric space E .

Proposition 1. *Let* T *be a finite subset of the compact metric space* E . *Then there exists a bounded projector on* $\mathcal{C}(E)$ *with the properties that*

(1) $$\mathcal{P}(P) \; = \; T \; ,$$

(2) $$P(1) \; = \; 1 \; .$$

Proof: Let

$$T \; = \; \{ a_1, \ldots, a_m \} \; .$$

We define the functions

$$w_1, \cdots, w_m \; \in \; \mathcal{C}(E)$$

by

(3) $$w_k(x) \; = \; d(x,a_1) \cdots d(x,a_m)/d(x,a_k) \qquad (1 \le k \le m) .$$

Then

$$w = w_1 + \cdots + w_m \in \mathscr{C}(E) \tag{4}$$

is a positive function. The functions

$$v_1, \ldots, v_m \in \mathscr{C}(E)$$

defined by

$$v_k = w_k / w \quad (1 \le k \le m) \tag{5}$$

satisfy the *duality relations*

$$v_k(a_l) = \delta_{k,l} \quad (1 \le k,l \le m) \ . \tag{6}$$

The operator P defined by

$$P(f) = \sum_{j=1}^{m} f(a_j)v_j \tag{7}$$

is a bounded projector on $\mathscr{C}(E)$. It follows from relation (6) that

$$P(f)(a_j) = f(a_j) \quad (1 \le j \le m) \tag{8}$$

is true for every $f \in \mathscr{C}(E)$. This implies

$$\mathscr{P}(P) = \{ a_1, \ldots, a_m \} \ . \tag{9}$$

Taking into account the relations (4) and (5) we obtain

$$P(1) = 1 \ . \tag{10}$$

This completes the proof of Proposition 1.

Remark 1. The projector P defined by (7) is called *Sheppard projector.*

A bounded projector P on $\mathscr{C}(E)$ is called *interpolation projector* if and only if the null space $\mathscr{N}(P)$ is an ideal and $P(1) = 1$.

Proposition 2. *The Sheppard projector is an interpolation projector.*

Proof: Relation (8) implies

(11) $$\mathcal{N}(P) = \mathcal{C}(E \| \{a_1, \dots, a_m\}) .$$

Taking into account (10) the proof of Proposition 2 is complete.

Proposition 3. *If* P *is an interpolation projector, then*

(12) $$\mathcal{P}(P) \neq \emptyset ,$$

(13) $$\mathcal{N}(P) = \mathcal{C}(E \| \mathcal{P}(P)).$$

Proof: Since P is a bounded linear projector on $\mathcal{C}(E)$, the ideal $\mathcal{N}(P)$ is closed. An application of Proposition 3 of Section 2.1 yields

(14) $$\mathcal{N}(P) = \mathcal{C}(E \| \mathcal{Z}(\mathcal{N}(P))) .$$

Since

$$\mathcal{N}(P) = \mathcal{R}(P^c)$$

we can conclude

$$\mathcal{Z}(\mathcal{N}(P)) = \mathcal{Z}(\mathcal{R}(P^c)) = \{ x \in E : f(x)-P(f)(x) = 0 \quad \forall f \in \mathcal{C}(E) \} = \mathcal{P}(P).$$

This implies relation (13).

It follows from $P(1) = 1$ that

$$\mathcal{N}(P) \neq \mathcal{C}(E) .$$

Since

$$\mathcal{C}(E \| T) = \mathcal{C}(E) \Leftrightarrow T = \emptyset ,$$

relation (12) follows. This completes the proof of Proposition 3.

Remark 2. If P is an interpolation projector, then

$$P^c(1) = 0 \quad .$$

Thus the remainder projector of an interpolation projector is not an interpolation projector.

2.3 Lattices of interpolation projectors

Let $\mathcal{L}$ be a set of commuting interpolation projectors on $\mathcal{C}(E)$ where E is a compact metric space. Then $\mathcal{L}$ generates a distributive lattice $\mathfrak{L}$ of commuting projectors on $\mathcal{C}(E)$. We will shown in this section that every projector from $\mathfrak{L}$ is an interpolation projector.

Proposition 1. *Let* P *and* Q *be commuting interpolation projectors on* $\mathcal{C}(E)$. *Then the product projector* PQ *is also an interpolation projector on* $\mathcal{C}(E)$. *Moreover the relation*

(1) $$\mathcal{P}(PQ) = \mathcal{P}(P) \cap \mathcal{P}(Q)$$

holds.

Proof: PQ is a bounded linear projector on $\mathcal{C}(E)$ satisfying

$$PQ(1) = 1 \quad .$$

It follows from Proposition 3 of Section 1.1 that

(2) $$\mathcal{N}(PQ) = \mathcal{N}(P) + \mathcal{N}(Q) \quad .$$

Thus $\mathcal{N}(PQ)$ is a closed ideal in $\mathcal{C}(E)$ and the product projector PQ is an interpolation projector on $\mathcal{C}(E)$.

It follows from Proposition 3 of Section 2.2 that

$$\mathcal{N}(P) = \mathcal{C}(E\|\mathcal{P}(P)) \;,\; \mathcal{N}(Q) = \mathcal{C}(E\|\mathcal{P}(Q)) \;,\; \mathcal{N}(PQ) = \mathcal{C}(E\|\mathcal{P}(PQ)) \;.$$

We apply Proposition 4 of Section 2.1 to relation (2) and obtain

$$\mathcal{C}(E\|\mathcal{P}(PQ)) = \mathcal{N}(PQ) = \mathcal{N}(P)+\mathcal{N}(Q) = \mathcal{C}(E\|\mathcal{P}(P))+\mathcal{C}(E\|\mathcal{P}(Q)) = \mathcal{C}(E\|\mathcal{P}(P)\cap\mathcal{P}(Q))$$

which implies $\mathcal{P}(PQ) = \mathcal{P}(P)\cap\mathcal{P}(Q)$. This completes the proof of Proposition 1.

Remark 1. It follows from Proposition 3 of Section 2.2 that the precision set of the interpolation projector PQ satisfies

$$\mathcal{P}(PQ) \quad = \quad \mathcal{P}(P) \cap \mathcal{P}(Q) \quad \neq \quad \emptyset \ .$$

Proposition 2. *Let* P *and* Q *be commuting interpolation projectors on* $\mathcal{C}(E)$. *Then the Boolean sum projector* $P \oplus Q$ *is also an interpolation projector on* $\mathcal{C}(E)$. *Moreover the relation*

$$\mathcal{P}(P \oplus Q) \quad = \quad \mathcal{P}(P) \cup \mathcal{P}(Q) \tag{3}$$

holds.

Proof: $P \oplus Q$ is a bounded linear projector on $\mathcal{C}(E)$ satisfying

$$P \oplus Q(\mathbf{1}) \quad = \quad \mathbf{1} \quad .$$

It follows from Proposition 4 of Section 1.1 that

$$\mathcal{N}(P \oplus Q) \quad = \quad \mathcal{N}(P) \cap \mathcal{N}(Q) \quad . \tag{4}$$

Thus $\mathcal{N}(P \oplus Q)$ is a closed ideal in $\mathcal{C}(E)$ and the Boolean sum projector $P \oplus Q$ is an interpolation projector on $\mathcal{C}(E)$.

It follows from Proposition 3 of Section 2.2 that

$$\mathcal{N}(P) = \mathcal{C}(E \| \mathcal{P}(P)) \ , \quad \mathcal{N}(Q) = \mathcal{C}(E \| \mathcal{P}(Q)) \ , \quad \mathcal{N}(P \oplus Q) = \mathcal{C}(E \| \mathcal{P}(P \oplus Q)) \quad .$$

We apply Proposition 2 of Section 2.1 to relation (4) and obtain

$$\mathcal{C}(E \| \mathcal{P}(P \oplus Q)) = \mathcal{N}(P \oplus Q) = \mathcal{N}(P) \cap \mathcal{N}(Q) = \mathcal{C}(E \| \mathcal{P}(P)) \cap \mathcal{C}(E \| \mathcal{P}(Q)) = \mathcal{C}(E \| \mathcal{P}(P) \cup \mathcal{P}(Q))$$

which implies

$$\mathcal{P}(P \oplus Q) \quad = \quad \mathcal{P}(P) \cup \mathcal{P}(Q) \quad .$$

This completes the proof of Proposition 2.

Proposition 3. *Let* $\mathcal{L}$ *be a set of commuting interpolation projectors on* $\mathcal{C}(E)$ *. Then the distributive lattice* $\mathfrak{L}$ *of commuting projectors generated from* $\mathcal{L}$ *consists of interpolation projectors.*

Proof: $\mathfrak{L}$ consists of all possible finite products and Boolean sums constructed with projectors from $\mathcal{L}$. It follows from Propositions 1 and 2 that these projectors are interpolation projectors. This completes the proof of Proposition 3.

Remark 2. Let $\mathfrak{L}_o$ denote the subset of projectors $P \in \mathfrak{L}$ with the property that $\mathcal{P}(P)$ is finite. If $\mathfrak{L}_o \neq \emptyset$, then $\mathfrak{L}_o$ is a sublattice of $\mathfrak{L}$.

Proposition 4. *Any interpolation projector* $P \in \mathfrak{L}$ *is uniquely determined by its precision set* $\mathcal{P}(P)$.

Proof: Assume that $P, Q \in \mathfrak{L}$ satisfy

$$\mathcal{P}(P) = T = \mathcal{P}(Q) .$$

Then we have

$$\mathcal{N}(P) = \mathcal{C}(E \| T) = \mathcal{N}(Q) .$$

Since

$$\mathcal{N}(P) = \mathcal{R}(I - P) \quad , \quad \mathcal{N}(Q) = \mathcal{R}(I - Q)$$

we obtain

$$Q(f) = QP(f) \quad , \quad P(f) = PQ(f) \quad (f \in \mathcal{C}(E)) .$$

Now $PQ = QP$ implies $P = Q$ which completes the proof of Proposition 4.

Notes, Section 2

The concept of precision set was introduced by GORDON - WIXOM [32] (see also GORDON - CHENEY [30])). An algebraic approach was used by DELVOS - SCHEMPP [23] to characterize interpolation projectors in spaces of continuous functions (see BIRKHOFF [9]). Our functional analytical presentation follows DELVOS - SCHEMPP [25]. NACHBIN [39] is a good source for functional analytical approximation theory (see also SEMADENI [42]) .

3 The method of parametric extension

3.1 The Dieudonné theorem

The Stone–Weierstraß approximation is a very powerful tool for the proof of general results on continuous functions by the device, which consists in proving these results for functions of a special type, and then extending them to all functions by a density argument. Here we consider as special functions the elements of the algebraic tensor product of two Banach algebras of continuous functions and we establish the Dieudonné theorem. This theorem will be constantly used in the method of parametric extension which is a basic tool for obtaining general results for bivariate continuous functions from univariate ones.

Let X and Y be compact metric spaces with distances d_1 and d_2, respectively. Then the cartesian product $X\times Y$ is again a compact metric space with distance

$$d((x,y),(u,v)) = d_1(x,u) + d_2(y,v) \qquad ((x,y) , (u,v) \in X\times Y) .$$

The *tensor product* $f\otimes g$ of the functions $f \in \mathcal{C}(X)$ and $g \in \mathcal{C}(Y)$ is defined by

$$f\otimes g(x,y) = f(x)g(y) \qquad (x \in X , y \in Y)$$

The *algebraic tensor product* of $\mathcal{C}(X)$ and $\mathcal{C}(Y)$ is defined by

$$\mathcal{C}(X)\otimes\mathcal{C}(Y) = \langle\{ f\otimes g : f \in \mathcal{C}(X) , g \in \mathcal{C}(Y) \}\rangle .$$

It is obvious that $\mathcal{C}(X)\otimes\mathcal{C}(Y)$ is a linear subspace of $\mathcal{C}(X\times Y)$.

Proposition 1. *$\mathcal{C}(X)\otimes\mathcal{C}(Y)$ is a subalgebra of $\mathcal{C}(X\times Y)$ which contains the constants.*

Proof: If $u, v \in \mathcal{C}(X)\otimes\mathcal{C}(Y)$, then

$$u = \sum_{i=1}^{m} f_i \otimes g_i \qquad (f_i, g_i \in \mathcal{C}(X) \ , \ 1 \le i \le m) \ ,$$

$$v = \sum_{j=1}^{n} p_j \otimes q_j \qquad (p_j, q_j \in \mathcal{C}(Y) \ , \ 1 \le j \le n) \ .$$

Then we have

$$u \otimes v = \sum_{i=1}^{m} \sum_{j=1}^{n} (f_i p_j) \otimes (g_i q_j)$$

which is an element of the algebraic tensor product $\mathcal{C}(X) \otimes \mathcal{C}(Y)$ .Moreover we have $1 \otimes 1 = 1$ which completes the proof of Proposition 1 .

Proposition 2. (Dieudonné) *The algebraic tensor product* $\mathcal{C}(X) \otimes \mathcal{C}(Y)$ *is dense in* $\mathcal{C}(X \times Y)$.

Proof: We apply the *Stone-Weierstrass theorem* to the algebra $\mathcal{C}(X) \otimes \mathcal{C}(Y)$. In view of Proposition 1 we have to show that the algebra $\mathcal{C}(X) \otimes \mathcal{C}(Y)$ separates the points of $X \times Y$. Let

$$(a,b) \ , \ (x,y) \in X \times X$$

with $(a,b) \neq (x,y)$. If $a \neq x$, $b \neq y$, then the function $d_1(\cdot,a) \otimes d_2(\cdot,b)$ separates the points (a,b) and (x,y) . If $a = x$, $b \neq y$, then the function $1 \otimes d_2(\cdot,b)$ separates (a,b) and (x,y) . If $a \neq x$, $b = y$, then the function $d_1(\cdot,a) \otimes 1$ separates (a,b) and (x,y) . This completes the proof of Proposition 2.

3.2 Tensor product operators

The method of parametric extension is a simple procedure for constructing linear operators on spaces of bivariate functions from linear operators which are defined on spaces of univariate functions. We use the Dieudonné theorem to show that parametric extensions of bounded linear operators are again bounded linear operators. By taking products of suitable parametric extensions we obtain the topological tensor product of bounded linear operators.

Assume that $f \in \mathscr{C}(X\times Y)$ and $x \in X$. We define $f^x \in \mathscr{C}(Y)$ by

$$f^x(t) = f(x,t) \quad (t \in Y) .$$

For $y \in Y$ we define ${}^y f \in \mathscr{C}(X)$ by

$${}^y f(s) = f(s,y) \quad (s \in X) .$$

Let A be a bounded linear operator on $\mathscr{C}(X)$. The *parametric extension* A' of A is defined by

$$A'(f)(x,y) = A({}^y f)(x) \quad .$$

Let B be a bounded linear operator on $\mathscr{C}(Y)$. The *parametric extension* B'' of B is defined by

$$B''(f)(x,y) = B(f^x)(y) \quad .$$

Proposition 1. *The restrictions* A' *and* B'' *to* $\mathscr{C}(X)\otimes\mathscr{C}(Y)$ *are linear operators on* $\mathscr{C}(X)\otimes\mathscr{C}(Y)$.

Proof: It is obvious that we have

(1) $$A'(f \otimes g) = Af \otimes g$$

and

(2) $$B''(f \otimes g) = f \otimes Bg$$

for $f \in \mathcal{C}(X)$ and $g \in \mathcal{C}(Y)$. The linearity of A' and B'' follows from their definitions. This completes the proof of Proposition 1.

Proposition 2. *The parametric extensions* A' *and* B'' *commute on* $\mathcal{C}(X)\otimes\mathcal{C}(Y)$.

Proof: It follows from the relations (1) and (2) that

(3) $$A'B''(f\otimes g) = Af\otimes Bg = B''A'(f\otimes g) \quad (f \in \mathcal{C}(X) , g \in \mathcal{C}(Y)).$$

This completes the proof of Proposition 2 .

The algebraic tensor product $\mathcal{C}(X)\otimes\mathcal{C}(Y)$ is a normed linear subspace of the Banach algebra $\mathcal{C}(X\times Y)$. We have

(4) $$\|f\otimes g\|_\infty = \|f\|_\infty \cdot \|g\|_\infty \quad (f \in \mathcal{C}(X) , g \in \mathcal{C}(Y)).$$

Proposition 3. *The parametric extensions* A' *and* B'' *are bounded linear operators on* $\mathcal{C}(X)\otimes\mathcal{C}(Y)$.

Proof: It is sufficient to consider the parametric extension A'. We assume that $h \in \mathcal{C}(X)\otimes\mathcal{C}(Y)$, $x \in X$, and $y \in Y$. Then we have

$$|A'(h)(x,y)| = |A({}^y h)(x)| \le \|A\| \cdot \|{}^y h\|_\infty .$$

Since

$$\|{}^y h\|_\infty \le \|h\|_\infty \quad (y \in Y)$$

we can conclude

(5) $$\|A'(h)\|_\infty \le \|A\|\cdot\|h\|_\infty \quad (h \in \mathcal{C}(X)\otimes\mathcal{C}(Y)) .$$

This completes the proof of Proposition 3.

Remark 1. It follows from the procf of Proposition 3 that the estimate (5) remains true for $h \in \mathcal{C}(X\times Y)$ in the following form:

(6) $$|A'(h)(x,y)| \le \|A\|\cdot\|h\|_\infty \quad (x \in X , y \in Y) .$$

In particular, $A'(h)$ is a bounded function on $X\times Y$.

Remark 2. Similar estimates hold for B'' :

(7) $$\|B''(h)\|_\infty \le \|B\|\cdot\|h\|_\infty \quad (h \in \mathcal{C}(X)\otimes\mathcal{C}(Y)) .$$

For $h \in \mathcal{C}(X\times Y)$ we have

(8) $$|B''(h)(x,y)| \le \|B\|\cdot\|h\|_\infty \quad (x \in X , x \in Y) .$$

Proposition 4. *The parametric extensions* A' *and* B'' *are bounded linear operators on* $\mathcal{C}(X\times Y)$.

Proof: It follows from the estimates (6) and (8) that it is sufficient to show that

(9) $$A'(h) , B''(h) \in \mathcal{C}(X\times Y) \quad (h \in \mathcal{C}(X\times Y)) .$$

It follows from the Dieudonné theorem that for every $h \in \mathcal{C}(X\times Y)$ there exist a sequence

$$h_n \in \mathcal{C}(X)\otimes\mathcal{C}(Y) \quad (n \in \mathbb{N})$$

with the property that

$$\lim_{n\to\infty} \|h - h_n\|_\infty = 0 . \tag{10}$$

We introduce the sequences

$$u_n = A'(h_n) \quad , \quad v_n = B''(h_n) \quad (n \in \mathbb{N}) .$$

It follows from Proposition 1 that

$$u_n , v_n \in \mathscr{C}(X)\otimes\mathscr{C}(Y) \quad (n \in \mathbb{N}) .$$

It follows from the estimates (5) and (7) that there exist functions $u, v \in \mathscr{C}(X\times Y)$ with the properties that

$$\lim_{n\to\infty} \|u - u_n\|_\infty = 0 , \tag{11}$$

$$\lim_{n\to\infty} \|v - v_n\|_\infty = 0 . \tag{12}$$

Taking into account the relations (6), (10), and (11) we obtain for every $(x,y) \in X\times Y$

$$\begin{aligned} &|u(x,y) - A'(h)(x,y)| \\ &\le |u(x,y) - u_n(x,y)| + |A'(h_n)(x,y) - A'(h)(x,y)| \\ &\le \|u - u_n\|_\infty + \|A\|\cdot\|h - h_n\|_\infty \end{aligned}$$

which implies

$$u = A'(h). \tag{13}$$

The equality

$$v = B''(h) \tag{14}$$

is verified similarly. This completes the proof of Proposition 4.

Proposition 5. (Fubini) *The parametric extensions* A' *and* B" *commute on* $\mathcal{C}(X \times Y)$.

Proof: In view of Proposition 4 the operator products A'B" and B"A' are bounded linear operators on $\mathcal{C}(X \times Y)$. It follows from Proposition 2 that

$$(15) \qquad A'B''(h) = B''A'(h) \qquad (h \in \mathcal{C}(X) \otimes \mathcal{C}(Y)) .$$

By the Dieudonné theorem relation (15) extends to all functions $h \in \mathcal{C}(X \times Y)$. This completes the proof of Proposition 5 .

Reamrk 3. In view of relation (3) A'B" is the algebraic tensor product $A \otimes B$ of A and B which acts on the tensor algebra $\mathcal{C}(X) \otimes \mathcal{C}(Y)$. In view of relation (15) the product A'B" is the *topological tensor product* of the bounded operators A and B being defined on $\mathcal{C}(X \times Y)$ which is also obtained in a unique way by the principle of extension of densely defined bounded linear operators.

3.3 Parametric extensions of projectors

In this section we apply the method of parametric extension to bounded projectors and in particular to interpolation projectors. It is shown that parametric extensions of interpolation projectors are again interpolation projectors. Due to the commutativity of parametrically extended operators we can apply the construction of distributive projector lattices and obtain as simple nontrivial instances the tensor product interpolation projector and the blending interpolation projector.

Let P be a bounded projector on $\mathcal{C}(X)$ and let Q be a bounded projector on $\mathcal{C}(Y)$.

Proposition 1. *The parametric extensions* P' *and* Q'' *are bounded projectors on* $\mathcal{C}(X\times Y)$ *which commute:*

(1) $$P'Q'' = Q''P' .$$

Proof: It follows from the results of Section 3.2 that P' and Q'' are bounded operators on $\mathcal{C}(X\times Y)$ which commute. Moreover we have for $f \in \mathcal{C}(X)$ and $g \in \mathcal{C}(Y)$:

(2) $$P'(f\otimes g) = Pf\otimes g ,$$

(3) $$Q''(f\otimes g) = f\otimes Qg .$$

This implies

(4) $$P'P'(h) = P'(h) \quad (h \in \mathcal{C}(X)\otimes\mathcal{C}(Y)) ,$$

(5) $$Q''Q''(h) = Q''(h) \quad (h \in \mathcal{C}(X)\otimes\mathcal{C}(Y)) .$$

Since P' and Q'' are bounded on $\mathcal{C}(X)\otimes\mathcal{C}(Y)$, an application of the Dieudonné theorem yields

(6) $$P'P' = P' .$$

(7) $$Q''Q'' = Q'' .$$

This completes the proof of Proposition 1.

Proposition 2. *Let* P *be an interpolation projector on* $\mathcal{C}(X)$ *and let* Q *be an interpolation projector on* $\mathcal{C}(Y)$. *Then the parametric extensions* P' *and* Q'' *are interpolation projectors on* $\mathcal{C}(X\times Y)$.

Proof: P' and Q'' are bounded projectors on $\mathcal{C}(X\times Y)$ satisfying

(8) $$P'(1) = 1 ,$$

(9) $$Q''(1) = 1 .$$

Recall that

$$\mathcal{N}(P) = \mathcal{C}(X \| \mathcal{P}(P)) , \quad \mathcal{N}(Q) = \mathcal{C}(Y \| \mathcal{P}(Q)) ,$$

and

$$P'(h)(x,y) = P({}^{y}h)(x) \quad (x \in X , \ h \in \mathcal{C}(X\times Y)) ,$$

$$Q''(h)(x,y) = Q(h^{x})(y) \quad (y \in Y , \ h \in \mathcal{C}(X\times Y)) .$$

This implies

(10) $$\mathcal{N}(P') = \mathcal{C}(X\times Y \| \mathcal{P}(P)\times Y),$$

(11) $$\mathcal{N}(Q'') = \mathcal{C}(X\times Y \| X\times\mathcal{P}(Q))$$

which completes the proof of Proposition 2.

Proposition 3. *The precision sets of the interpolation projectors* P' *and* Q" *are given by*

$$\mathcal{P}(P') = \mathcal{P}(P)\times Y \quad , \tag{12}$$

$$\mathcal{P}(Q'') = X\times\mathcal{P}(Q) \quad . \tag{13}$$

Proof: Recall that P' and Q" are interpolation projectors. We apply Proposition 3 of Section 2.2. to the relations (10) and (11) and obtain (12) and (13). This completes the proof of Proposition 3.

Proposition 4. *The product projector* P'Q" *is an interpolation projector on* $\mathcal{C}(X\times Y)$. *Its precision set is given by*

$$\mathcal{P}(P'Q'') = \mathcal{P}(P)\times\mathcal{P}(Q) \quad . \tag{14}$$

Proof: It follows from Proposition 1 that the interpolation projectors P' and Q" commute. Taking into account Proposition 1 of Section 2.3 it follows that P'Q" is an interpolation projector with the property that

$$\mathcal{P}(P'Q'') = \mathcal{P}(P') \cap \mathcal{P}(Q'') \quad .$$

Since

$$\mathcal{P}(P') = \mathcal{P}(P)\times Y \quad , \quad \mathcal{P}(Q'') = X\times\mathcal{P}(Q) \quad ,$$

relation (14) follows. This completes the proof of Proposition 4.

Example 1. We consider interpolation projectors with finite precision sets:

$$\mathcal{P}(P) = \{x_1,\dots, x_m\} \quad , \quad \mathcal{P}(Q) = \{y_1,\dots, y_n\} \quad .$$

Then we have

$$\mathcal{R}(P) = \langle f_1,\ldots, f_m\rangle \;, \quad \mathcal{R}(Q) = \langle g_1,\ldots, g_n\rangle$$

with

$$f_i(x_j) = \delta_{i,j} \quad (1 \le i,j \le m) \;, \quad g_k(y_l) = \delta_{k,l} \quad (1 \le k,l \le n) \;,$$

$$f_1 + \cdots + f_m = 1 \;, \quad g_1 + \cdots + g_n = 1 \;.$$

The interpolation projectors P and Q are given by

$$P(f) = \sum_{i=1}^{m} f(x_i)f_i \qquad (f \in \mathcal{C}(X))$$

$$Q(g) = \sum_{j=1}^{n} g(y_j)g_j \qquad (g \in \mathcal{C}(Y)) \;,$$

Then the parametric extensions P' and Q'' are given by

$$P'(h)(x,y) = \sum_{i=1}^{m} h(x_i,y)f_i(x) \qquad (h \in \mathcal{C}(X\times Y)) \;,$$

$$Q''(h)(x,y) = \sum_{j=1}^{n} h(x,y_j)g_j(y) \qquad (h \in \mathcal{C}(X\times Y)) \;.$$

Thus the interpolation projector $P'Q''$ possesses the representation

(15) $$P'Q''(h)(x,y) = \sum_{i=1}^{m} \sum_{j=1}^{n} h(x_i,y_j)f_i(x)g_j(y) \qquad (h \in \mathcal{C}(X\times Y)) \;.$$

$P'Q''$ is also called *tensor product interpolation projector* . Its precision set is given by

(16) $$\mathcal{P}(P'Q'') = \{ (x_i,y_j) : 1 \le i \le m,\ 1 \le j \le n \} \;.$$

Proposition 5. *The Boolean sum projector* $P'\oplus Q''$ *is an interpolation projector on* $\mathcal{C}(X\times Y)$. *Its precision set is given by*

$$\mathcal{P}(P'\oplus Q'') = (\mathcal{P}(P)\times Y) \cup (X\times\mathcal{P}(Q)) \quad . \tag{17}$$

Proof: It follows from Proposition 2 of Section 2.3 that $P'\oplus Q''$ is an interpolation projector on $\mathcal{C}(X\times Y)$. Moreover we have

$$\mathcal{P}(P'\oplus Q'') = \mathcal{P}(P')\cup\mathcal{P}(Q) = (\mathcal{P}(P)\times Y)\cup(X\times\mathcal{P}(Q)) \quad .$$

This completes the proof of Proposition 5 .

Example 2. We continue Example 1 .The interpolation projector $P'\oplus Q''$ possesses the representation

$$P'\oplus Q''(h)(x,y) = \sum_{i=1}^{m} h(x_i,y)f_i(x) + \sum_{j=1}^{n} h(x,y_j)g_j(y) - \sum_{i=1}^{m}\sum_{j=1}^{n} h(x_i,y_j)f_i(x)g_j(y)$$

where $h \in \mathcal{C}(X\times Y)$. $P'\oplus Q''$ is also called *blending interpolation projector*. Its precision set is given by

$$\mathcal{P}(P'\oplus Q'') = \{ (x_i,y) : 1\le i\le m, y\in Y\} \cup \{ (x,y_j) : x\in X, 1\le j\le n\} \ .$$

Notes, Section 3

The method of parametric extension is the basic tool in extending univariate results of the theory of interpolation and approximation to the bivariate case (see GORDON [29] and GORDON - CHENEY [30]). Our representation is based on the Dieudonné theorem and follows DELVOS - SCHEMPP [24] . It enables an easy approach to tensor product operators on spaces of continuous functions on product spaces. For a more general treatment we refer to LIGHT - CHENEY [37].

4 Projectors and right inverses

4.1 Construction of right invertible operators

Right invertible operators are used to describe additional properties of elements of a linear space. These properties are required in the formulation of remainder representations of interpolation and approximation processes. We use projectors and invertible operators which are defined on the whole linear space and which satisfy a null condition to construct a right invertible operator on an appropriate subspace.

Let X be a linear space and R be an injective linear operator on X, i. e.,

$$\mathcal{N}(R) = \{0\} . \tag{1}$$

Let X_o be a linear subspace of X with the property that

$$\mathcal{R}(R) \subseteq X_o . \tag{2}$$

We assume that P is a projector on X_o such that

$$PR = 0 . \tag{3}$$

It is easily seen that

$$\mathcal{R}(R) \cap \mathcal{R}(P) = \{ 0 \} . \tag{4}$$

Moreover

$$X_1 := \mathcal{R}(R) + \mathcal{R}(P) \tag{5}$$

is a subspace of X_o.

Proposition 1. *There is a unique linear operator* A *in* X *with domain* $\mathcal{D}(A) = X_1$ *such that*

$$f = P(f) + RA(f) \quad (f \in \mathcal{D}(A)) . \tag{6}$$

Proof: Let

$$f = u + v \quad (u \in \mathcal{R}(P) , v \in \mathcal{R}(R)) .$$

Using relation (3) we obtain

$$P(f) = u \quad , \quad f - P(f) = v .$$

Since

$$v = R(g) \quad (g \in X) \quad ,$$

it follows from relation (1) that

$$g = A(f)$$

is the unique solution of the *defining equation* (6) of A . The linearity of A is a consequence of relation (1) and the linearity of relation (6) .

Proposition 2. *The operator* A *possesses* R *as a right inverse:*

(7) $$AR = I .$$

Proof: If $f = R(g)$, then relation (3) yields

$$f = RA(f) = RAR(g) \quad ,$$

i. e.,

$$R(g) = RAR(g) \quad (g \in X) \quad .$$

Taking into account (1) we obtain

$$g = AR(g) \quad (g \in X)$$

which completes the proof of Proposition 2.

Proposition 3. *The operator* $Q = RA$ *is a projector on* X_1 *with the property that*

$$\mathcal{R}(Q) = \mathcal{R}(R). \tag{8}$$

Proof: Taking into account (7) we obtain

$$QQ = RARA = RA = Q .$$

It is obvious that

$$\mathcal{R}(Q) \subseteq \mathcal{R}(R) .$$

If $f = R(g)$, it follows from (7) that

$$Q(f) = RAR(g) = Rg = f$$

which implies

$$\mathcal{R}(Q) \supseteq \mathcal{R}(R) .$$

This completes the proof of Proposition 3.

Proposition 4. *The relation*

$$\mathcal{N}(A) = \mathcal{R}(P) \tag{9}$$

holds.

Proof: This follows from (6) and (1) .

Remark 1. P is called the *initial projector* of A . In view of (9) $\mathcal{R}(P)$ is called the *space of constants* of the right invertible operator A.

Example 1. Let $X = \mathcal{C}(\mathbb{R}_+)$. R is the operator of definite integration :

$$R(f)(x) = \int_0^x f(s)\,ds \qquad (f \in \mathscr{C}(\mathbb{R}_+)) .$$

The projector P is defined on $X_0 = X$:

$$P(f)(x) = f(0) .$$

Then we have

$$X_1 = \mathscr{C}^1(\mathbb{R}_+) , \quad A(f) = Df = \frac{df}{dx} ,$$

$$Q(f)(x) = \int_0^x Df(s)\,ds = f(x) - f(0) .$$

4.2 The abstract Taylor formula

The right invertible operator A defined by P and R generates a sequence X_m of linear subspaces :

$$X_m = \mathcal{D}(A^m) \quad (m \in \mathbb{N}) .$$

Proposition 1. *The operators*

$$R^j PA^j \quad (0 \le j \le n)$$

are projectors on X_n .

Proof: Note that

$$X_1 \supseteq X_2 \supseteq \cdots \supseteq X_n .$$

Using Proposition 2 of Section 4.1 we obtain

$$(R^j PA^j)(R^j PA^j) = R^j PA^j R^j PA^j = R^j PPA^j = R^j PA^j .$$

This completes the proof of Proposition 1.

Proposition 2. *The projectors* $R^j PA^j$ $(0 \le j \le n)$ *commute and are orthogonal on* X_n .

Proof: Since $PR = 0$ and $AR = I$, we obtain for $j < k$:

$$(R^j PA^j)(R^k PA^k) = R^j P A^j R^j PA^k = R^j PA^j R^j R^{k-j} PA^k = R^j P R^{k-j} PA^k = 0 ,$$

i. e.,

$$(R^j PA^j)(R^k PA^k) = 0 \quad (0 \le j < k \le n) .$$

Taking into account relation Proposition 4 of Section 4.1 we get

$$(R^kPA^k)(R^jPA^j) = R^kPA^{k-j}A^jR^jPA^j = R^kPA^{k-j}PA^j = 0$$

which implies

$$(R^kPA^k)(R^jPA^j) = 0 \quad (0 \le j < k \le n) .$$

This completes the proof of Proposition 2.

Proposition 3. *The operator*

$$P_m = \sum_{j=0}^{m-1} R^jPA^j$$

is a projector on X_{m-1} $(m \in \mathbb{N})$.

Proof: It follows from Proposition 2 that the Boolean sum

$$R^0PA^0 \oplus R^1PA^1 \oplus \cdots \oplus R^{m-1}PA^{m-1}$$

is a projector on X_{m-1} . Since the projectors

$$R^0PA^0,\ R^1PA^1,\ \ldots,\ R^{m-1}PA^{m-1}$$

are orthogonal, we have

$$R^0PA^0 \oplus R^1PA^1 \oplus \cdots \oplus R^{m-1}PA^{m-1} = P_m .$$

This completes the proof of Proposition 3.

Since R is injective, the linear operator R^m is also injective:

(1) $$\mathcal{N}(R^m) = \{ 0 \} \quad (m \in \mathbb{N}) .$$

It follows from Proposition 2 of Section 4.1 that the relation

(2) $$\mathcal{R}(R^m) \subseteq X_{m-1} \quad (m \in \mathbb{N})$$

holds.

Proposition 4. *The relation*

$$P_m R^m = 0$$

is true for every $m \in \mathbb{N}$.

Proof: Using (3) and (7) of Section 4.1 we obtain

$$P_m R^m = \sum_{j=0}^{m-1} R^j P A^j R^m = \sum_{j=0}^{m-1} R^j P R^{m-j} = 0 .$$

This completes the proof of Proposition 4.

Proposition 5. (Taylor) A^m *is the unique linear operator in* X *which is defined by the injective operator* R^m *and the projector* P_m :

(4) $$f = \sum_{j=0}^{m-1} R^j P A^j(f) + R^m A^m(f) \quad (f \in X_m) .$$

Proof: We apply Proposition 1 of Section 4.1 to the operators R^m and P_m with

$$\mathcal{D}(P_m) = X_{m-1} .$$

We denote by B the unique right invertible operator with right inverse R^m and initial projector P_m . Then we have

$$f = P_m(f) + R^m B(f) , \quad \mathcal{D}(B) = \mathcal{R}(R^m) + \mathcal{R}(P_m).$$

It follows from (3), (7), and (9) of Section 4.1 that

$$\mathcal{D}(B) \subseteq \mathcal{D}(A^m) \quad .$$

In particular we have

(5) $$A^m P_m = 0 \quad ,$$

(6) $$A^m R^m = I \quad .$$

It follows from the defining equation of A that

$$A^j(f) = PA^j(f) + RA^{j+1}(f) \quad (f \in \mathcal{D}(A^m) \ , \ 0 \le j < m) \ .$$

Then we can conclude

$$f = P(f) + RA(f) = P(f) + RPA(f) + R^2A^2(f) = \sum_{j=0}^{m-1} R^jPA^j(f) + R^mA^m(f) \ ,$$

i. e.,

(8) $$f = P_m(f) + R^mA^m(f) \quad .$$

This shows $f \in \mathcal{D}(B)$ and $B = A^m$ which completes the proof of Proposition 5.

Proposition 6. *The operator*

$$Q_m = R^mA^m$$

is a projector on X_m .

Proof: This follows from Proposition 3 of Section 4.1 and Proposition 5.

Example 1. We continue Example 1 of Section 4.1 . Recall that

$$P(f)(x) = f(0) \quad , \quad R(f)(x) = \int_0^x f(s)\,ds \quad , \quad A(f) = Df \quad .$$

It is easily seen that

$$R^j(1)(x) = \frac{x^j}{j!} \quad ,$$

$$R^jPA^j(f)(x) = D^jf(0)\cdot\frac{x^j}{j!} \quad .$$

Let

$$h_m(x) = \int_0^x \frac{(x-s)^{m-1}}{(m-1)!} f(s)\, ds \quad .$$

We have

$$h_1(x) = R(f)(x) \quad .$$

Proceeding by induction we have for $m > 1$:

$$h_m(x) = \int_0^x \frac{(x-s)^{m-1}}{(m-1)!} DR(f)(s)\, ds = \frac{(x-s)^{m-1}}{(m-1)!} R(f)(s)\Big|_0^x + \int_0^x \frac{(x-s)^{m-2}}{(m-2)!} R(f)(s)\, ds \quad ,$$

i. e.,

$$h_m(x) = \int_0^x \frac{(x-s)^{m-2}}{(m-2)!} R(f)(s)\, ds \quad .$$

Thus we have shown

$$R^m(f)(x) = \int_0^x \frac{(x-s)^{m-1}}{(m-1)!} f(s)\, ds \quad .$$

An application of Proposition 5 yields the *classical Taylor formula* :

$$f(x) = \sum_{j=0}^{m-1} D^jf(0)\cdot\frac{x^j}{j!} + \int_0^x \frac{(x-s)^{m-1}}{(m-1)!} D^mf(s)\, ds \qquad (f \in \mathcal{C}^m(\mathbb{R}_+)) \quad .$$

4.3 Parametric extensions of right invertible operators

The properties of parametric extensions of bounded operators make it possible to construct parametric extension of right invertible operators. The commutativity of parametrically extended bounded and right invertible operators is established. These results are of importance in remainder formulas.

Let X and Y be compact metric spaces. Let P be a bounded projector on $\mathfrak{C}(X)$ an R be a bounded injective operator on $\mathfrak{C}(X)$ such that

$$PR = 0 . \tag{1}$$

Then there is a unique right invertible operator A in $\mathfrak{C}(X)$ with right inverse R and initial projector P . It follows from Proposition 1 of Section 3.3 that P' is a bounded projector on $\mathfrak{C}(X\times Y)$.

Proposition 1. *The parametric extension* R' *is a bounded linear and injective operator on* $\mathfrak{C}(X\times Y)$ *with the property that*

$$P'R' = 0 . \tag{2}$$

Proof: It follows from Proposition 4 of Section 3.2 that R' is a bounded linear operator on $\mathfrak{C}(X\times Y)$. Since

$$R'(f)(x,y) = R(^{y}f)(x) \qquad (x \in X , y \in Y) ,$$

the relation

$$\mathcal{N}(R) = \{ 0 \}$$

implies

$$\mathcal{N}(R') = \{ 0 \} . \tag{3}$$

Moreover we have for $f \in \mathfrak{C}(X)$ and $g \in \mathfrak{C}(Y)$:

$$P'R'(f\otimes g) = (PRf)\otimes g = 0 .$$

An application of the Dieudonné theorem yields

$$P'R' = 0$$

which completes the proof of Proposition 1.

Proposition 2. *There is a unique right invertible operator* A' *with right inverse* R' *and initial projector* P' .

Proof: In view of (2) and (3) we can apply Proposition 1 of Section 4.1 and obtain the existence of A'. This completes the proof of Proposition 2.

A' is called the *parametric extension* of the right invertible operator A . This is justified by the following

Proposition 3. *If* $h \in \mathcal{C}(X \times Y)$ *, then*

$$h \in \mathcal{D}(A') \quad \Leftrightarrow \quad {}^{y}h \in \mathcal{D}(A) \quad (\forall\, y \in Y) \ .$$

Moreover the relation

$$(4) \qquad A'(h)(x,y) = A({}^{y}h)(x,y) \quad (x \in X \ , \ y \in Y) \ .$$

holds.

Proof: Recall that

$$h = P'(h) + R'A'(h) \quad (h \in \mathcal{D}(A')) \ .$$

For every $x \in X$ and $y \in Y$ we have

$$h(x,y) = P'(h)(x,y) + R'(A'(h))(x,y) \quad \Leftrightarrow$$

$${}^{y}h(x) - P({}^{y}h)(x) = R({}^{y}(A'(h)))(x) \quad \Leftrightarrow$$

$${}^{y}h(x) - P({}^{y}h)(x) = R(A({}^{y}h))(x) \quad \Leftrightarrow$$

$${}^{y}(A'(h))(x) \quad = \quad A({}^{y}h)(x)$$

whence (4) follows . This completes the proof of Proposition 3.

Prposition 4. *If* $f \in \mathcal{D}(A)$ *and* $g \in \mathcal{C}(Y)$, *then* $f{\otimes}g \in \mathcal{D}(A')$ *and*

$$A'(f{\otimes}g) \quad = \quad (Af){\otimes}g \quad . \tag{5}$$

Proof: Relation (6) follows from Proposition 3.

Let V be a bounded linear projector on $\mathcal{C}(Y)$. It follows from Proposition 4 of Section 3.2 that V'' is a bounded linear operator on $\mathcal{C}(X{\times}Y)$. The following result may be considered as an extension of Proposition 5 (*Fubini*) of Section 3.2 .

Proposition 5. (Leibniz) *The parametric extensions* A' *and* V'' *commute on* $\mathcal{D}(A')$:

$$A'V''(h) \quad = \quad V''A'(h) \quad (h \in \mathcal{D}(A')) \quad . \tag{6}$$

Proof: Recall that

$$h - P'(h) \quad = \quad R'A''(h) \quad (h \in \mathcal{D}(A')) \quad .$$

Applying Proposition 5 of Section 3.2 we obtain

$$R'V''A'(h) \quad = \quad V''R'A'(h) \quad = \quad V''(h-P'(h))$$

$$= \quad V''(h)-V''P'(h) \quad = \quad V''(h)-P'V''(h) \quad = \quad R'A'(V''(h))$$

which implies

$$V''(h) \in \mathcal{D}(A') \quad , \quad A'V''(h) \quad = \quad V''A'(h) \quad (h \in \mathcal{D}(A')) \quad .$$

This completes the proof of Proposition 5.

Proposition 6. *If* $f \in \mathcal{D}(A)$ *and* $g \in \mathcal{C}(Y)$, *then* $f \otimes g \in \mathcal{D}(A')$ *and*

$$A'V''(f \otimes g) = (Af) \otimes (Vg) \quad . \tag{7}$$

Proof: Recall that

$$V''(f \otimes g) = f \otimes (Vg) \quad .$$

Now an application of Proposition 4 yields (7) which completes the proof of Proposition 6 .

Example 1. Let

$$X = [a,b] \quad , \quad Y = [c,d]$$

be compact intervals in $\mathbb{R}$. We choose P and R as in Example 1 of Section 4.1 :

$$P(f)(x) = f(a) \quad ,$$

$$R(f)(x) = \int_a^x f(s)\, ds \quad (f \in \mathcal{C}([a,b])) \ .$$

Then A is given by

$$A(f)(x) = Df(x) \quad (f \in \mathcal{C}^1([a,b]) = \mathcal{D}(A)) \ .$$

The parametric extensions P' , R' , and A' are given by:

$$P'(h)(x,y) = h(a,y) \quad ,$$

$$R'(h)(x,y) = \int_a^x h(s,y)\, ds \quad ,$$

$$A'(h) = D_1 h = \frac{\partial h}{\partial x}$$

with

$$\mathcal{D}(A') = \mathcal{C}^{1,0}([a,b]\times[c,d]) .$$

The operator V is defined by

$$V(g)(y) = \int_c^d g(t)\,dt \qquad (g \in \mathcal{C}([c,d])) ,$$

i. e., V is a bounded linear functional on $\mathcal{C}([c,d])$. Since $1 \in \mathcal{C}([c,d])$, V is also a bounded linear operator on $\mathcal{C}([c,d])$. We apply Proposition 5 to the operators A and V and obtain the *classical Leibniz theorem* :

$$\eta(x) = \int_c^d h(x,t)\,dt ,$$

$$D\eta(x) = \int_c^d D_1 h(x,t)\,dt \qquad (h \in \mathcal{C}^{1,0}([a,b]\times[c,d])) .$$

All considerations which have been carried out for $\mathcal{C}(X)$ remain true for $\mathcal{C}(Y)$. For later applications we list the corresponding results. Let Q be a bounded projector on $\mathcal{C}(Y)$ and S be a bounded injective linear operator on $\mathcal{C}(Y)$ such that

(8) $$QS = 0 .$$

Let B denote the right invertible operator in $\mathcal{C}(Y)$ with right inverse S and initial projector Q . Then Q" is a bounded projector on $\mathcal{C}(X\times Y)$ and S" is a bounded injective linear operator on $\mathcal{C}(X\times Y)$ with the property that

(9) $$Q''S'' = 0 .$$

B" is the unique right invertible operator in $\mathcal{C}(X\times Y)$ with right inverse S" and initial projector Q" .

Let U be a bounded linear operator on $\mathcal{C}(X)$. Then the parametric extensions U' and B" commute on $\mathcal{D}(B'')$.

Example 2. We extend Example 1 to the second variable. We have

$$Q(g)(x) = g(c) ,$$

$$S(g)(y) = \int_c^y g(t)\,dt \quad (g \in \mathscr{C}([c,d])) ,$$

$$B(g) = Dg .$$

The parametric extensions are given by

$$Q''(h)(x,y) = h(x,c) ,$$

$$S''(h)(x,y) = \int_c^y h(x,t)\,dt \quad (g \in \mathscr{C}([c,d])) ,$$

$$B''(h) = D_2 h = \frac{\partial h}{\partial y}$$

with

$$h \in \mathscr{C}^{0,1}([a,b]\times[c,d]) = \mathcal{D}(B'') .$$

The bounded linear operator U on $\mathscr{C}([a,b])$ is defined by

$$U(f)(x) = \int_a^b f(t)\,dt \quad (f \in \mathscr{C}([a,b])) .$$

Then we have

$$\gamma(y) = \int_a^b h(s,y)\,ds ,$$

$$D\gamma(y) = \int_a^b D_2 h(s,y)\,ds \quad (h \in \mathscr{C}^{0,1}([a,b]\times[c,d])).$$

4.4 Tensor product of right invertible operators

In this section we combine the methods of Boolean sum and parametric extensions to construct tensor products of right invertible operators.

It follows from Proposition 1 of Section 3.3 that the bounded projectors P' and Q'' commute on $\mathscr{C}(X\times Y)$. Its Boolean sum

$$P'\oplus Q'' = P' + Q'' - P'Q''$$

is a bounded projector on $\mathscr{C}(X\times Y)$. Since R' and S'' are bounded injective linear operators on $\mathscr{C}(X\times Y)$, its product

$$(1) \qquad R'S'' = S''R'$$

is also a bounded injective linear operator on $\mathscr{C}(X\times Y)$.

Proposition 1. *The relation*

$$(2) \qquad (P'\oplus Q'')(R'S'') = 0$$

holds.

Proof: Taking into account (2) and (10) in Section 4.3 we obtain

$$(P'\oplus Q'')(R'S'') = P'R'S'' + Q''R'S'' - P'Q''R'S'' = Q''S''R' - P'R'Q''S'' = 0$$

which completes the proof of Proposition 1 .

In view of the results of Section 4.1 there exists a unique right invertible operator T with right inverse $R'S''$ and initial projector $P'\oplus Q''$. We have

$$(3) \qquad h = P'(h) + Q''(h) - P'Q''(h) + R'S''T(h) ,$$

$$(4) \qquad \mathcal{D}(T) = \mathcal{R}(P'\oplus Q'') + \mathcal{R}(R'S'') .$$

Recall that

(5) $$\mathcal{R}(P'\oplus Q'') \quad = \quad \mathcal{R}(P') + \mathcal{R}(Q'') \quad .$$

We introduce the linear subspace $\mathcal{D}_0(T)$ of $\mathcal{D}(T)$:

$$\mathcal{D}_0(T) \quad = \quad \{\ h \in \mathcal{D}(T) : P'(h) \in \mathcal{D}(B'') \ ,\ Q''(h) \in \mathcal{D}(A')\ \}\ .$$

Proposition 2. *The inclusion*

(6) $$\mathcal{D}(A)\otimes\mathcal{D}(B) \quad \subseteq \quad \mathcal{D}_0(T)$$

holds. The restriction of T *to* $\mathcal{D}(A)\otimes\mathcal{D}(B)$ *is the algebraic tensor product of the operators* A *and* B *:*

(7) $$T(f\otimes g) \quad = \quad A(f)\otimes B(g) \quad = \quad A'B''(f\otimes g) \quad (\ f \in \mathcal{D}(A)\ ,\ g \in \mathcal{D}(B)\ ) \quad .$$

Proof: We verify the defining equation (3) of T :

$$\begin{aligned} & f\otimes g \quad - \quad (P'\otimes Q'')(f\otimes g) \\ = \quad & f\otimes g \quad - \quad P(f)\otimes g \quad - \quad f\otimes Q(g) + P(f)\otimes Q(g) \\ = \quad & (f - P(f))\otimes g \quad - \quad (f - P(f))\otimes Q(g) \\ = \quad & RA(f)\otimes g \quad - RA(f)\otimes Q(g) \\ = \quad & RA(f)\otimes(g - Q(g)) \\ = \quad & RA(f)\otimes SB(g) \\ = \quad & R'S''(A(f)\otimes B(g)) \end{aligned}$$

whence (7) follows . This completes the proof of Proposition 2.

Proposition 3. *The inclusion*

(8) $$\mathcal{R}(R'S'') \subseteq \mathcal{D}_0(T)$$

is true.

ProoF: This follows from $R'S'' = S''R'$ and

$$P'R' = 0 \quad , \quad Q''S'' = 0$$

which completes the proof of Proposition 3.

Proposition 4. *If* $h \in \mathcal{D}_0(T)$, *then*

$$P' \oplus Q''(h) \in \mathcal{D}_0(T) \quad ,$$

i. e., the restriction $(P' \oplus Q'')_0$ of $P' \oplus Q''$ *to* $\mathcal{D}_0(T)$ *is a projector on* $\mathcal{D}_0(T)$.

Proof: Assume that

$$h = P' \oplus Q''(u) \quad (u \in \mathcal{D}_0(T)) \quad .$$

Then we have

$$P'(h) = P'(u) \in \mathcal{D}(B'') \quad , \quad Q''(h) = Q''(u) \in \mathcal{D}(A') \quad .$$

Since $\mathcal{R}(P' \oplus Q'') \subseteq \mathcal{D}(T)$, we obtain $h \in \mathcal{D}_0(T)$. This completes the proof of Proposition 4.

Proposition 5. *The relation*

(9) $$\mathcal{D}_0(T) = \{ h \in \mathcal{D}(A') \cap \mathcal{D}(B'') : A'(h) \in \mathcal{D}(B'') \}$$

is true.

Proof: Recall that

$$\mathcal{D}(A') = \mathcal{R}(P') + \mathcal{R}(R') \quad , \quad \mathcal{D}(B") = \mathcal{R}(Q") + \mathcal{R}(S") \quad .$$

Since $\mathcal{D}_0(T) \subseteq \mathcal{D}(T)$, we have for $h \in \mathcal{D}_0(T)$:

$$h = P'(h) + Q"(h) - P'Q"(h) + R'S"T(h) \quad ,$$

$$P'(h) \in \mathcal{D}(B") \quad , \quad Q"(h) \in \mathcal{D}(A') \quad .$$

This implies $h \in \mathcal{D}(A')\cap\mathcal{D}(B")$. Moreover we can conclude

$$A'(h) = A'Q"(h) + S"T(h) = Q"A'(h) + S"T(h)$$

which implies $A'(h) \in \mathcal{D}(B")$.

Assume now that

$$h \in \mathcal{D}(A')\cap\mathcal{D}(B") \quad , \quad A'(h) \in \mathcal{D}(B") \quad .$$

Then we have

$$\begin{aligned} h &= P'(h) + R'A'(h) \\ &= P'(h) + R'Q"A'(h) + R'S"B"A'(h) \\ &= P'(h) + R'A'Q"(h) + R'S"B"A'(h) \\ &= P'(h) + Q"(h) - P'Q"(h) + R'S"B"A'(h) \end{aligned}$$

which implies $h \in \mathcal{D}(T)$ and

(10) $$T(h) = B"A'(h) \quad .$$

Since

$$P'Q" = Q"P' \quad , \quad R'S" = S"R' \quad ,$$

we get

$$P'(h) \in \mathcal{D}(B") \quad , \quad Q"(h) \in \mathcal{D}(A') \quad .$$

Thus we have shown

$$h \in \mathcal{D}_0(T)$$

which completes the proof of Proposition 5 .

The following result may be considered as an extension of Proposition 5 (*Fubini*) of Section 3.2 and of Proposition 5 (*Leibniz*) of Section 4.3.

Proposition 6. (Schwarz) *Suppose that*

$$h \in \mathcal{D}(A')\cap\mathcal{D}(B'') \quad , \quad A'(h) \in \mathcal{D}(B'') \quad .$$

Then

$$B''(h) \in \mathcal{D}(A')$$

and

$$A'B''(h) = B''A'(h), \tag{11}$$

i. e., the parametric extensions A' *and* B" *commute on*

$$\mathcal{D}_0(T) = \{ h \in \mathcal{D}(A')\cap\mathcal{D}(B'') : A'(h) \in \mathcal{D}(B'') \}.$$

Proof: We have

$$B''(h) - P'B''(h) = B''(h) - B''P'(h) = B''R'A'(h) = R'B''A'(h)$$

which implies $B''(h) \in \mathcal{D}(A')$ and

$$A'B''(h) = B''A'(h) \quad .$$

This completes the proof of Proposition 6.

Remark 1. It follows from Propositions 3, 4, and 5 and the results of Section 4.1 that the *tensor product operator* A'B" of the right invertible operators A' and B" can be defined as the unique right invertible operator with *right inverse* R'S" and *initial projector* $(P'\oplus Q'')_0$. The domain of A'B" is given by

(12) $\mathcal{D}(A'B'') = \mathcal{D}_0(T) = \{ h \in \mathcal{D}(A') \cap \mathcal{D}(B'') : A'(h) \in \mathcal{I}(B'') \}$.

The tensor product operator $A'B''$ is the restriction of the right invertible operator T defined by $R'S''$ and $P' \oplus Q''$ on the linear subspace $\mathcal{D}_0(T)$ of $\mathcal{D}(T)$.

Example 1. We continue the example of Section 4.3. We have

$$A'B''(h) = D_1 D_2 h = \frac{\partial^2 h}{\partial x \partial y} \qquad (h \in \mathcal{C}^{1,1}([a,b]\times[c,d])) .$$

An application of Proposition 6 yields the *classical Schwarz rule*:

$$D_1 D_2 h = D_2 D_1 h \qquad (h \in \mathcal{C}^{1,1}([a,b]\times[c,d])) .$$

4.5 A bivariate Taylor formula

The technique of Boolean projector and the commutativity of parametric extensions of bounded operators and right invertible operators are used to establish an abstract Taylor formula for functions in two variables.

Let A be a right invertible operator in $\mathcal{C}(X)$ with right inverse R and initial projector P. Then A' is the right invertible operator in $\mathcal{C}(X\times Y)$ with right inverse R' and initial projector P'. For $f \in \mathcal{D}(A'^m)$ we have in view of Proposition 5 of Section 4.2

$$f = \sum_{j=0}^{m-1} R'^j P' A'^j (f) + R'^m A'^m (f) .$$

The operator

$$P'_m = \sum_{j=0}^{m-1} R'^j P' A'^j \tag{1}$$

is a projector on $\mathcal{D}(A'^m)$ with remainder projector

$$I - P'_m = R'^m A'^m . \tag{2}$$

Let B be a right invertible operator in $\mathcal{C}(Y)$ with right inverse S and initial projector Q. As before it follows that

$$Q''_m = \sum_{k=0}^{m-1} S''^k Q'' B''^k \tag{3}$$

is a projector on $\mathcal{D}(B''^m)$ with remainder projector

$$I - Q''_m = S''^m B''^m . \tag{4}$$

Recall that $A'B''$ is a right invertible operator in $\mathcal{C}(X\times Y)$ with right inverse $R'S''$ and initial projector $(P'\oplus Q'')_0$ being defined on

$$\mathcal{D}(A'B'') = \{ h \in \mathcal{D}(A')\cap\mathcal{D}(B'') : A'(h) \in \mathcal{D}(B'') \} .$$

The operator

$$(R'S")^m(A'B")^m \quad .$$

is a projector on $\mathcal{D}((A'B")^m)$. Its remainder projector is given by

(6) $$I - (R'S")^m(A'B")^m = \sum_{j=0}^{m-1} (R'S")^j(P'\oplus Q")(A'B")^j \; .$$

Proposition 1. *The relation*

(7) $$P'_m\oplus Q"_m = \sum_{j=0}^{m-1} (R'S")^j(P'\oplus Q")(A'B")^j$$

holds.

Proof: We apply Proposition 5 (Fubini) of Section 3.2, Proposition 5 (Leibniz) of Section 4.3, and Proposition 6 (Schwarz) of Section 4.4 . Taking into account the relations (2), (4), and (6) we obtain

$$\begin{aligned}
& f - \sum_{j=0}^{m-1} (R'S")^j(P'\oplus Q")(A'B")^j(f) \\
&= (R'S")^m(A'B")^m(f) \\
&= R'^m S"^m A'^m B"^m(f) \\
&= R'^m A'^m S"^m B"^m(f) \\
&= R'^m A'^m(f - Q"_m(f)) \\
&= f - Q"_m(f) - P'_m(f - Q"_m(f)) \\
&= f - (P'_m\oplus Q"_m)(f)
\end{aligned}$$

whence relation (7) follows. This proves Proposition 1.

Remark 1. We have shown that

$$(A'B'')^m = A'^m B''^m$$

is a right invertible operator with right inverse

$$(R'S'')^m = R'^m S''^m$$

and initial projector

$$(8)\quad P'_m \oplus Q''_m = \sum_{j=0}^{m-1} R'^j P' A'^j + \sum_{k=0}^{m-1} S''^k Q'' B''^k - \sum_{j=0}^{m-1} \sum_{k=0}^{m-1} R'^j S''^k P' Q'' A'^j B''^k .$$

Proposition 2. (Bivariate Taylor formula) *The equation*

$$(9)\quad f = \sum_{j=0}^{m-1} R'^j P' A'^j(f) + \sum_{k=0}^{m-1} S''^k Q'' B''^k(f) - \sum_{j=0}^{m-1} \sum_{k=0}^{m-1} R'^j S''^k P' Q'' A'^j B''^k(f)$$

$$+ \quad (R'S'')^m (A'B'')^m (f)$$

is true for every $f \in \mathcal{D}((A'B'')^m)$.

Proof: Relation (9) follows from (8) ,(7) , and (6) . This completes the proof of Proposition 2.

Example 1. We continue Example 1 of Section 4.2 and Example 1 of Section 4.4 . We have

$$\mathcal{D}((D_1D_2)^m) = \mathscr{C}^{m,m}([a,b]\times[c,d]) \quad .$$

An application of Proposition 2 yields the *classical bivariate Taylor formula* for $f \in \mathscr{C}^{m,m}([a,b]\times[c,d])$:

$$f(x,y) = \sum_{j=0}^{m-1} D_1^j f(a,y)\cdot\frac{(x-a)^j}{j!} + \sum_{k=0}^{m-1} D_2^k f(x,c)\cdot\frac{(y-c)^k}{k!}$$

$$- \sum_{j=0}^{m-1}\sum_{k=0}^{m-1} D_1^j D_2^k f(a,c)\cdot\frac{(x-a)^j}{j!}\cdot\frac{(y-c)^k}{k!}$$

$$+ \int_a^x\int_c^y \frac{(x-s)^{m-1}}{(m-1)!}\cdot\frac{(y-t)^{m-1}}{(m-1)!}\, D_1^m D_2^m f(s,t)\, ds\, dt \quad .$$

Notes, Section 4

The algebraic theory of right invertible operators is an important tool in constructing remainder projectors and new interpolation projectors. We refer to PRZEWORSKA-ROLEWICZ [41] and TASCHE [45]. The application to the bivariate case follows DELVOS - SCHEMPP [24] (see also BASZENSKI [1] and BASZENSKI - DELVOS [3]).

5 Boolean bivariate polynomial interpolation

5.1 Biermann interpolation

We are going to apply the preceding methods to bivariate polynomial interpolattion. In particular, we investigate interpolation projectors which are obtained from chains of parametric extensions of univariate interpolation projectors. The space of interpolants is determined via Boolean methods and an explicit formula for the Lagrange fundamental functions is derived with the aid of the sum representation for bivariate chains of interpolation projectors.

The univariate interpolation projectors

$$P_1, \dots, P_r, \quad Q_1, \dots, Q_r$$

are polynomial interpolation projectors:

$$P_m(f)(x) = \sum_{i=1}^{k_m} f(x_i)\phi_{i,m}(x) \quad , \quad Q_n(g)(y) = \sum_{j=1}^{l_n} g(y_j)\psi_{j,n}(y) \; .$$

The interpolation points satisfy

$$\{ x_1, \dots, x_{k_m} \} \subseteq [a,b] \quad , \quad \{ y_1, \dots, y_{l_n} \} \subseteq [c,d]$$

with

(1) $$1 \le k_1 < k_2 < \dots < k_r \quad , \quad 1 \le l_1 < l_2 < \dots < l_r \; .$$

The cardinal functions are given by

$$\phi_{i,m}(x) = \prod_{\substack{k=1 \\ k \ne i}}^{k_m} \frac{(x - x_k)}{(x_i - x_k)} \quad (1 \le i \le k_m) \quad , \quad \psi_{j,n}(y) = \prod_{\substack{l=1 \\ l \ne j}}^{l_n} \frac{(y - y_l)}{(y_j - y_l)} \quad (1 \le j \le l_n).$$

Then we have

(2) $\mathcal{R}(P_m) = \langle 1, x, \ldots, x^{k_m - 1} \rangle = \Pi_{k_m - 1}$, $\mathcal{R}(Q_n) = \langle 1, y, \ldots, y^{l_n - 1} \rangle = \Pi_{l_n - 1}$.

It follows from (1) and (2) that the parametric extensions

$$P'_1, \ldots, P'_r, \quad Q''_1, \ldots, Q''_r$$

are bivariate interpolation projectors which form chains :

(3) $$P'_1 \le \ldots \le P'_r \quad , \quad Q''_1 \le \ldots \le Q''_r \quad .$$

Moreover we have

(4) $$P'_m Q''_n \quad = \quad Q''_n P'_m \qquad (1 \le m,n \le r) \quad .$$

$P'_m Q''_n$ is the projector of *bivariate polynomial tensor product interpolation* . We have the representation

(5) $$P'_m Q''_n(f)(x,y)) \quad = \quad \sum_{i=1}^{k_m} \sum_{j=1}^{l_n} f(x_i, y_j)\ \phi_{i,m}(x) \cdot \psi_{j,n}(y) \ .$$

$P'_m Q''_n$ possesses the interpolation properties

(6) $$P'_m Q''_n(f)(x_i, y_j) \quad = \quad f(x_i, y_j) \qquad (1 \le i \le k_m \ , \ 1 \le j \le l_n \ ) \ .$$

The space of interpolants defined by $P'_m Q''_n$ is

(7) $$\mathcal{R}(P'_m Q''_n) \quad = \quad \Pi_{k_m - 1} \otimes \Pi_{l_n - 1} \quad .$$

The projectors

$$P'_1 \ , \ \ldots \ , \ P'_r \ , \ Q''_1 \ , \ \ldots, \ Q''_r \quad .$$

generate a distributive lattice $\mathfrak{L}$ of interpolation projectors on $\mathcal{C}([a,b] \times [c,d])$. The projector B_r of *Biermann interpolation* is a special element of $\mathfrak{L}$ which is defined by

$$B_r = P'_1Q''_r \oplus P'_2Q''_{r-1} \oplus \cdots \oplus P'_rQ''_1$$

where $r \in \mathbb{N}$. It follows from Proposition 1 of Section 1.3 and relation (3) that the projectors $B_1,\dots, B_r$ form a finite chain :

$$B_1 \le B_2 \le \cdots \le B_r \quad .$$

Proposition 1. *The space of interpolants defined by the Biermann interpolation projector* B_r *is given by*

$$\mathcal{R}(B_r) = \Pi_{k_1-1} \otimes \Pi_{l_r-1} + \cdots + \Pi_{k_r-1} \otimes \Pi_{l_1-1} \quad . \tag{8}$$

Proof: It follows from Proposition 4 of Section 1.1 that

$$\mathcal{R}(B_r) = \mathcal{R}(P'_1Q''_r) + \cdots + \mathcal{R}(P'_rQ''_1) \quad .$$

Applying (7) to this relation we obtain (8). This completes the proof of Proposition 1.

Proposition 2. *The Biermann interpolation projector satisfies the interpolation conditions*

$$B_r(f)(x_i,y_j) = f(x_i,y_j) \quad (1 \le i \le k_m,\ 1 \le j \le l_{r+1-m},\ 1 \le m \le r). \tag{9}$$

Proof: Taking into account Proposition 2 of Section 2.3 we obtain

$$\mathcal{P}(B_r) = \mathcal{P}(P'_1Q''_r) \cup \cdots \cup \mathcal{P}(P'_rQ''_1) \quad . \tag{10}$$

Since

$$(11)\qquad \mathcal{P}(P'_m Q''_n) = \{ (x_i,y_j) : 1\le i\le k_m ,\ 1\le j\le l_n \} \qquad (m+n = r+1 ,\ 1\le m\le r) ,$$

relation (9) follows. This completes the proof of Proposition 2.

It follows from (1), (10), and (11) that the set of interpolation points of the Biermann interpolation projector possesses the *disjoint representation*

$$(12)\qquad \mathcal{P}(B_r) = \bigcup_{m=1}^{r} \bigcup_{n=0}^{r-m} \{ (x_i,y_j) : k_{m-1} < i \le k_m ,\ l_{r-m-n} < j \le l_{r+1-m-n} \}$$

with $k_0 := 0$, $l_0 := 0$.

The number of interpolation points of an interpolation projector with finite precision set is given by the dimension of the range of the interpolation projector. This is used to determine the cardinality of the precision set of the Biermann interpolation projector B_r .

Proposition 3. *The number of interpolation points of the Biermann interpolation projector satisfies the relation*

$$(13)\qquad |\mathcal{P}(B_r)| = \sum_{m=1}^{r} k_m (l_{r+1-m} - l_{r-m})$$

with $l_0 = 0$.

Proof: We apply Proposition 2 of Section 1.3 and obtain the *sum representation* of the Biermann interpolation projector:

$$(14)\qquad B_r = P'_1 Q''_r + \cdots + P'_r Q''_1 - (P'_1 Q''_{r-1} + \cdots + P'_{r-1} Q''_1) .$$

This implies the *orthogonal sum representation* of the Biermann interpolation projector:

$$(15)\qquad B_r = \sum_{m=1}^{r} P'_m (Q''_{r+1-m} - Q''_{r-m}) \qquad (Q''_0 := 0) .$$

Then we can conclude

$$
\begin{aligned}
&|\mathcal{P}(B_r)| \\
&= \dim(\mathcal{R}(B_r)) \\
&= \sum_{m=1}^{r} \dim(\mathcal{R}(P'_m(Q''_{r+1-m} - Q''_{r-m}))) \\
&= \sum_{m=1}^{r} \dim(\mathcal{R}(P'_m Q''_{r+1-m})) - \sum_{m=1}^{r} \dim(\mathcal{R}(P'_m Q''_{r-m})) \\
&= \sum_{m=1}^{r} k_m l_{r+1-m} - \sum_{m=1}^{r} k_m l_{r-m}
\end{aligned}
$$

whence (13) follows . This completes the proof of Proposition 3.

Using the disjoint representation (12) of the precision set $\mathcal{P}(B_r)$ of the Biermann interpolation projector we obtain the *Lagrange representation* of the Biermann interpolant :

$$
(16) \qquad B_r(f) = \sum_{m=1}^{r} \sum_{n=0}^{r-m} \sum_{i=1+k_{m-1}}^{k_m} \sum_{j=1+l_{r-m-n}}^{l_{r+1-m-n}} f(x_i, y_j)\, \Phi_{i,j} .
$$

Proposition 4. *The cardinal functions of Biermann interpolation are given by*

$$
(17) \qquad \Phi_{i,j}(x,y) = \sum_{s=m}^{m+n} \phi_{i,s}(x)\cdot\psi_{j,r+1-s}(y) - \sum_{s=m}^{m+n-1} \phi_{i,s}(x)\cdot\psi_{j,r-s}(y)
$$

$$
(k_{m-1} < i \le k_m,\ l_{r-m-n} < j \le l_{r+1-m-n},\ 0 \le n \le r - m,\ 1 \le m \le r).
$$

Proof: We apply the Biermann interpolation projector B_r to the function

$$
f(x,y) = \phi_{i,r}(x)\cdot\psi_{j,r}(y)
$$

and obtain

$$B_r(\phi_{i,r} \otimes \psi_{j,r}) = \Phi_{i,j} .$$

Using the sum representation (14) of the Biermann interpolation projector we can conclude

$$\begin{aligned} \Phi_{i,j} &= \sum_{s=1}^{r} P_s(\phi_{i,r}) \otimes Q_{r+1-s}(\psi_{j,r}) - \sum_{s=1}^{r-1} P_s(\phi_{i,r}) \otimes Q_{r-s}(\psi_{j,r}) \\ &= \sum_{s=m}^{r} \phi_{i,s} \otimes Q_{r+1-s}(\psi_{j,r}) - \sum_{s=m}^{r-1} \phi_{i,s} \otimes Q_{r-s}(\psi_{j,r}) \\ &= \sum_{s=m}^{m+n} \phi_{i,s} \otimes \psi_{j,r+1-s} - \sum_{s=m}^{m+n-1} \phi_{i,s} \otimes \psi_{j,r-s} . \end{aligned}$$

This completes the proof of Proposition 4.

5.2 Triangular elements

The bivariate polynomial interpolation of functions given on a regular *triangular mesh* was studied by Biermann . We call the projector B_r Biermann interpolation projector since it enables the construction of the classical Biermann interpolation interpolant on a triangular mesh as a special case . Polynomial interpolation on a triangular mesh is a basic tool in the finite element method ([48]).

We choose the interpolation points

$$x_i = \frac{i \cdot h}{r} \ , \quad y_j = \frac{j \cdot h}{r} \quad (1 \le i,j \le r \ , \ h > 0 \ ) ,$$

$$k_m = m \quad , \quad l_n = n \quad (\ 1 \le m,n \le r \ ) \ .$$

The univariate interpolation projectors are given by

$$P_m(f)(x) = \sum_{i=1}^{m} f(x_i)\phi_{i,m}(x) \quad , \quad Q_n(g)(y) = \sum_{j=1}^{n} g(y_j)\psi_{j,n}(y) \ .$$

with

$$\phi_{i,m}(x) = \prod_{\substack{k=1 \\ k \ne i}}^{m} \frac{(x - x_k)}{(x_i - x_k)} \quad (1 \le i \le m) \ , \quad \psi_{j,n}(y) = \prod_{\substack{l=1 \\ l \ne j}}^{n} \frac{(y - y_l)}{(y_j - y_l)} \quad (1 \le j \le n).$$

Then we have

(1) $$\mathcal{R}(P'_m Q''_n) = \Pi_{m-1} \otimes \Pi_{n-1}$$

and

(2) $$\mathcal{P}(P'_m Q''_n) = \{ \ (x_i,y_j) : 1 \le i \le m \ , \ 1 \le j \le n \ \} \ .$$

Recall that the Biermann interpolation projector is given by

$$B_r = P'_1 Q''_r \oplus P'_2 Q''_{r-1} \oplus \cdots \oplus P'_r Q''_1 \ .$$

The set of interpolation points of B_r is obtained from relation (10) of Section 5.1 and relation (2) :

$$\mathcal{P}(B_r) = \bigcup_{m=1}^{r} \mathcal{P}(P'_m Q''_{r+1-m}) = \bigcup_{m=1}^{r} \{ (x_i, y_j) : 1 \le i \le m \ , \ 1 \le j \le r+1-m \} \ ,$$

i. e., we have

$$\text{(3)} \qquad \mathcal{P}(B_r) = \bigcup_{m=1}^{r} \{ (\frac{i \cdot h}{r}, \frac{j \cdot h}{r}) : 1 \le i \le m \ , \ 1 \le j \le r+1-m \}$$

which is a regular triangular mesh. The space of interpolants defined by the classical Biermann projector B_r is obtained from relation (8) of Section 5.1 :

$$\text{(4)} \qquad \mathcal{R}(B_r) = \Pi_0 \otimes \Pi_{r-1} + \cdots + \Pi_{r-1} \otimes \Pi_0 \ ,$$

i. e., $\mathcal{R}(B_r)$ is the space of *complete bivariate polynomials of degree* $r-1$:

$$\text{(5)} \qquad \mathcal{R}(B_r) = \langle\{ x^i y^j : 0 \le i+j \le r-1 \ , \ i,j \ge 0 \}\rangle \ .$$

We use Proposition 3 of Section 5.1 to determine the number of interpolation points :

$$\text{(6)} \qquad |\mathcal{P}(B_r)| = \sum_{m=1}^{r} m = \binom{r+1}{2} \ .$$

The Lagrange representation of the classical Biermann interpolation projector is obtained from relation (16) of Section 5.1 :

$$\text{(7)} \qquad B_r(f) = \sum_{m=1}^{r} \sum_{n=0}^{r-m} f(x_m, y_{r+1-m-n}) \ \Phi_{m,r+1-m-n} \ .$$

The cardinal functions of classical Biermann interpolation are given by relation (17) of Section 5.1 :

(8)
$$\Phi_{m,r+1-m-n}(x,y)$$
$$= \sum_{s=m}^{m+n} \phi_{m,s}(x)\cdot\psi_{r+1-m-n,r+1-s}(y) - \sum_{s=m}^{m+n-1} \phi_{m,s}(x)\cdot\psi_{r+1-m-n,r-s}(y)$$
$$(0 \le n \le r - m , 1 \le m \le r) .$$

Example 1. We present a list of the cardinal functions for the case $r = 3$:

$$(x_1,y_3) : \quad \Phi_{1,3}(x,y) = \phi_{1,1}(x)\cdot\psi_{3,3}(y) ,$$
$$(x_2,y_2) : \quad \Phi_{2,2}(x,y) = \phi_{2,2}(x)\cdot\psi_{2,2}(y) ,$$
$$(x_3,y_1) : \quad \Phi_{3,1}(x,y) = \phi_{3,3}(x)\cdot\psi_{1,1}(y) ,$$

$$(x_1,y_2) : \quad \Phi_{1,2}(x,y) = \phi_{1,1}(x)\cdot\psi_{2,3}(y) + \phi_{1,2}(x)\cdot\psi_{2,2}(y) - \phi_{1,1}(x)\cdot\psi_{2,2}(y) ,$$

$$(x_2,y_1) : \quad \Phi_{2,1}(x,y) = \phi_{2,2}(x)\cdot\psi_{1,2}(y) + \phi_{2,3}(x)\cdot\psi_{1,1}(y) - \phi_{2,2}(x)\cdot\psi_{1,1}(y) ,$$

$$(x_1,y_1) : \quad \Phi_{1,1}(x,y) = \phi_{1,1}(x)\cdot\psi_{1,3}(y) + \phi_{1,2}(x)\cdot\psi_{1,2}(y) + \phi_{1,3}(x)\cdot\psi_{1,1}(y) - \phi_{1,1}(x)\cdot\psi_{1,2}(y) - \phi_{1,2}(x)\cdot\psi_{1,1}(y) .$$

5.3 Rectangular elements .

In this section we use the Biermann interpolation projector to construct a family of bivariate polynomial interpolation methods which are important in the finite element analysis of plate bending problems.

We specify the set of interpolation points :

$$x_{2i-1} = -\frac{(1+2(r-i))\cdot h}{(2r-1)\cdot 2} \quad , \quad x_{2i} = -x_{2i-1},$$

$$y_{2j-1} = -\frac{(1+2(r-j))\cdot h}{(2r-1)\cdot 2} \quad , \quad y_{2j} = -y_{2j-1}$$

$$(\ 1 \le i,j \le r\ )\ .$$

Moreover we choose

$$k_m = 2m \ , \ l_n = 2n \quad (\ 1 \le m,n \le r\ )\ .$$

The univariate interpolation projectors are given by

$$P_m(f)(x) = \sum_{i=1}^{2m} f(x_i)\phi_{i,m}(x) \quad , \quad Q_n(g)(y) = \sum_{j=1}^{2n} g(y_j)\psi_{j,n}(y).$$

with

$$\phi_{i,m}(x) = \prod_{\substack{k=1\\k\neq i}}^{2m} \frac{(x - x_k)}{(x_i - x_k)} \quad (1 \le i \le 2m)\ , \quad \psi_{j,n}(y) = \prod_{\substack{l=1\\l\neq j}}^{2n} \frac{(y - y_l)}{(y_j - y_l)} \quad (1 \le j \le 2n).$$

Then we have

(1) $$\mathcal{R}(P'_m Q''_n) = \Pi_{2m-1} \otimes \Pi_{2n-1}$$

and

(2) $$\mathcal{P}(P'_m Q''_n) = \{\ (x_i,y_j) : 1 \le i \le 2m\ ,\ 1 \le j \le 2n\ \}\ .$$

The Biermann interpolation projector is given by

$$B_r = P'_1 Q''_r \oplus P'_2 Q''_{r-1} \oplus \cdots \oplus P'_r Q''_1\ .$$

It follows from (1) and relation (8) of Section 5.1 that the space of interpolants defined by B_r is given by

(3) $$\mathcal{R}(B_r) = \Pi_1 \otimes \Pi_{2r-1} + \cdots + \Pi_{2r-1} \otimes \Pi_1 .$$

It follows from (3) that $\mathcal{R}(B_r)$ contains the space of complete bivariate polynomials of degree 2r-1 :

$$\langle\{ x^i y^j : 0 \le i+j \le 2r-1 , i,j \ge 0 \} \subseteq \mathcal{R}(B_r) .$$

The set of interpolation points is obtained from (2) and relation (10) of Section 5.1 :

(4) $$\mathcal{P}(B_r) = \bigcup_{m=1}^{r} \{ (x_i,y_j) : 1 \le i \le 2m , 1 \le j \le 2(r+1-m) \}.$$

To compute the number of interpolation points of B_r we apply Proposition 3 of Section 5.1 .

(5) $$|\mathcal{P}(B_r)| = \sum_{m=1}^{r} 2m\cdot 2 = 4\binom{r+1}{2} .$$

The Lagrange representation of the *rectangular Biermann interpolant* is obtained from relation (16) of Section 5.1 :

$$B_r(f) = \sum_{m=1}^{r} \sum_{n=0}^{r-m} \sum_{i=1+2m-1}^{2m} \sum_{j=2(r-m-n)-1}^{2(r-m-n)+2} f(x_i,y_j)\, \Phi_{i,j}$$

where

$$\Phi_{i,j}(x,y) = \sum_{s=m}^{m+n} \phi_{i,s}(x)\cdot\psi_{j,r+1-s}(y) - \sum_{s=m}^{m+n-1} \phi_{i,s}(x)\cdot\psi_{j,r-s}(y)$$

$$(2m-2 < i \le 2m,\ 2(r-m-n) < j \le 2(r-m-n)+2,\ 0 \le n \le r-m,\ 1 \le m \le r).$$

Example 1. We illustrate the method of rectangular Biermann interpolation for the case $r = 2$ which yields a well known *serendipity element* of the finite element method [48]:

$$(x_1,y_3) = (\frac{-h}{2},\frac{-h}{3\cdot 2}) : \quad \Phi_{1,3}(x,y) = \phi_{1,1}(x)\cdot\psi_{3,2}(y) ,$$

$$(x_1,y_4) = (\frac{-h}{2},\frac{+h}{3\cdot 2}) : \quad \Phi_{1,4}(x,y) = \phi_{1,1}(x)\cdot\psi_{4,2}(y) ,$$

$$(x_2,y_3) = (\frac{+h}{2},\frac{-h}{3\cdot 2}) : \quad \Phi_{2,3}(x,y) = \phi_{2,1}(x)\cdot\psi_{3,2}(y) ,$$

$$(x_2,y_4) = (\frac{+h}{2},\frac{+h}{3\cdot 2}) : \quad \Phi_{2,4}(x,y) = \phi_{2,1}(x)\cdot\psi_{4,2}(y) ,$$

$$(x_3,y_1) = (\frac{-h}{3\cdot 2},\frac{-h}{2}) : \quad \Phi_{3,1}(x,y) = \phi_{3,2}(x)\cdot\psi_{1,1}(y) ,$$

$$(x_4,y_1) = (\frac{+h}{3\cdot 2},\frac{-h}{2}) : \quad \Phi_{4,1}(x,y) = \phi_{4,2}(x)\cdot\psi_{1,1}(y) ,$$

$$(x_3,y_2) = (\frac{-h}{3\cdot 2},\frac{+h}{2}) : \quad \Phi_{3,2}(x,y) = \phi_{3,2}(x)\cdot\psi_{2,1}(y) ,$$

$$(x_4,y_2) = (\frac{+h}{3\cdot 2},\frac{+h}{2}) : \quad \Phi_{4,2}(x,y) = \phi_{4,2}(x)\cdot\psi_{2,1}(y) ,$$

$$(x_1,y_1) = (\frac{-h}{2},\frac{-h}{2}) :$$

$$\Phi_{1,1}(x,y) = \phi_{1,1}(x)\psi_{1,2}(y) + \phi_{1,2}(x)\psi_{1,1}(y) - \phi_{1,1}(x)\psi_{1,1}(y) ,$$

$$(x_1,y_2) = (\frac{-h}{2},\frac{+h}{2}) :$$

$$\Phi_{1,2}(x,y) = \phi_{1,1}(x)\psi_{2,2}(y) + \phi_{1,2}(x)\psi_{2,1}(y) - \phi_{1,1}(x)\psi_{2,1}(y) ,$$

$$(x_2,y_1) = (\frac{+h}{2},\frac{-h}{2}) :$$

$$\Phi_{2,1}(x,y) = \phi_{2,1}(x)\psi_{1,2}(y) + \phi_{2,2}(x)\psi_{1,1}(y) - \phi_{2,1}(x)\psi_{1,1}(y) ,$$

$$(x_2,y_2) = (\frac{+h}{2},\frac{+h}{2}) :$$

$$\Phi_{2,2}(x,y) = \phi_{2,1}(x)\psi_{2,2}(y) + \phi_{2,2}(x)\psi_{2,1}(y) - \phi_{2,1}(x)\psi_{2,1}(y) .$$

5.4 The univariate polynomial interpolation error

We review the different remainder formulas for univariate polynomial Lagrange interpolation. In particular the integral representation of the interpolation error will be used in bivariate interpolation.

Let the interpolation points $x_1, \dots, x_k$ satisfy

$$a \le x_i \le b \quad , \quad b = a + h \quad (1 \le i \le k \; , \; h > 0) \; .$$

We denote the associated polynomial Lagrange interpolation projector by

$$P_k(f)(x) \quad = \quad \sum_{i=1}^{k} f(x_i) \prod_{\substack{j=1 \\ j \ne i}}^{k} \frac{(x - x_j)}{(x_i - x_j)} \; .$$

Then we have the representation formula

$$(1) \qquad f(x) \quad = \quad P_k(f)(x) \quad + \quad [x, x_1, \dots, x_k](f) \cdot (x - x_1) \cdots (x - x_k) \; .$$

Recall that the divided differences $[x_0, x_1, \dots, x_k](f)$ is given by

$$[x_0, x_1, \dots, x_k](f) \quad = \quad \sum_{i=0}^{k} f(x_i) \prod_{\substack{j=0 \\ j \ne i}}^{k} \frac{1}{(x_i - x_j)} \; .$$

Assume that $f \in \mathscr{C}^k([a,b])$. The Taylor formula yields the decomposition

$$f(x) \quad = \quad T_k(f)(x) + R_k(f)(x)$$

with

$$T_k(f)(x) \quad = \quad \sum_{i=0}^{k-1} D^i f(a) \frac{(x-a)^i}{i!} \quad , \quad R_k(f)(x) \quad = \quad \int_a^b \frac{(x-a)_+^{k-1}}{(k-1)!} D^k f(t) \, dt \; .$$

Since

$$[x, x_1, \dots, x_k](T_k(f)) \quad = \quad 0 \; ,$$

we obtain from relation (1) that

$$\begin{aligned} & f(x) - P_k(f)(x) \\ &= [x,x_1,\ldots,x_k](f)(x-x_1)\cdots(x-x_k) \\ &= [x,x_1,\ldots,x_k](R_k(f))(x-x_1)\cdots(x-x_k) \ , \end{aligned}$$

i. e. we have

(2) $$f(x) - P_k(f)(x) = \int_a^b G_k(x,y)\cdot D^k f(t)\, dt.$$

The *kernel* $G_k(x,t)$ of the remainder of polynomial Lagrange interpolation is given by

(3) $$G_k(x,t) = M(x,x_1,\ldots,x_k,t)\cdot(x-x_1)\cdots(x-x_k)$$

with

(4) $$M(x_0,x_1,\ldots,x_k,t) = \sum_{i=0}^{k} \left(\prod_{\substack{j=0 \\ j\neq i}}^{k} \frac{1}{(x_i-x_j)} \right) \cdot \frac{(x_i-t)_+^{k-1}}{(k-1)!} \ .$$

The function $M(x_0,x_1,\ldots,x_k,t)$ is the B-spline of degree $k-1$ with knots $x_0, x_1,\ldots, x_k$. We list some properties of the B-spline $M(x_0,x_1,\ldots,x_k,t)$ as a function of t. Let

$$\alpha = \min\{x_0,x_1,\ldots,x_k\} \ , \quad \beta = \max\{x_0,x_1,\ldots,x_k\} \ .$$

Then we have

$$M(x_0,x_1,\ldots,x_k,t) = 0 \quad (t < \alpha,\ t > \beta),$$

$$M(x_0,x_1,\ldots,x_k,t) > 0 \quad (\alpha < t < \beta),$$

$$\int_\alpha^\beta M(x_0,x_1,\ldots,x_k,t)\, dt = \frac{1}{k!} \ .$$

Using these properties of the B-spline we obtain the remainder of polynomial Lagrange interpolation in *Cauchy form* :

$$(5)\quad f(x) - P_k(f)(x) = (x-x_1)\cdots(x-x_k)\frac{D^k f(\xi)}{k!} \quad (a \le x \le b,\ a < \xi < b).$$

Recall that $b = a+h$. It follows from (5) that the *order of convergence* of approximation of polynomial Lagrange interpolation via P_k is k, i. e., we have

$$(6)\quad f(x) - P_k(f)(x) = \mathcal{O}(h^k) \quad (h \to 0) .$$

5.5 The error in polynomial blending and tensor product interpolation

We assume that the interpolation points satisfy

$$a \le x_i \le b \quad , \quad b = a + h \qquad (1 \le i \le k \; , \; h > 0) \; ,$$

$$c \le y_j \le d \quad , \quad d = c + h \qquad (1 \le j \le l \; , \; h > 0) \; .$$

The univariate interpolation projectors are given by

$$P_k(f)(x) \quad \sum_{i=1}^{k} f(x_i) \prod_{\substack{j=1 \\ j \neq i}}^{k} \frac{(x - x_j)}{(x_i - x_j)} \quad , \quad Q_l(g)(y) = \sum_{j=1}^{l} g(y_j) \prod_{\substack{r=1 \\ r \neq j}}^{l} \frac{(y - y_r)}{(y_j - y_r)} \; .$$

The associated remainder projectors are

$$P_k^c(f)(x) = f(x) - P_k(f)(x) = \int_a^b G_k(x,s) \cdot D^k f(s) \, ds \qquad (f \in \mathcal{C}^k([a,b])) \; ,$$

$$Q_l^c(g)(y) = g(y) - Q_l(g)(y) = \int_c^d H_l(y,t) \cdot D^l g(t) \, dt \qquad (g \in \mathcal{C}^l([c,d])) \; .$$

The remainder kernels satisfy

(2)
$$G_k(x,s) = M(x,x_1,..,x_k,s)(x-x_1)\cdot\cdot(x-x_k),$$

$$H_l(y,t) = M(y,y_1,..,y_l,t)(y-y_1)\cdot\cdot(y-y_l).$$

We will determine the remainder of *bivariate polynomial blending interpolation* . Its interpolation projector is given by

$$P'_k \oplus Q''_l(f)(x,y) = \sum_{i=1}^{k} f(x_i,y) \prod_{\substack{j=1 \\ j \neq i}}^{k} \frac{(x - x_j)}{(x_i - x_j)} + \sum_{j=1}^{l} f(x,y_j) \prod_{\substack{r=1 \\ r \neq j}}^{l} \frac{(y - y_r)}{(y_j - y_r)}$$

$$- \sum_{i=1}^{k} \sum_{j=1}^{l} f(x_i,y_j) \cdot \prod_{\substack{j=1 \\ j \neq i}}^{k} \frac{(x - x_j)}{(x_i - x_j)} \cdot \prod_{\substack{r=1 \\ r \neq j}}^{l} \frac{(y - y_r)}{(y_j - y_r)}$$

The remainder projectors P'^{c}_k and Q''^{c}_l commute on $\mathcal{C}^{k,l}([a,b]\times[c,d])$. It follows from relation (4) of Section 1.1 that the equation

$$(P'_k \oplus Q''_l)^c = P'^{\,c}_k \cdot Q''^{\,c}_l$$

holds. Taking into account relation (1) we obtain the *remainder formula of bivariate polynomial blending interpolation* :

$$(3) \qquad f(x,y) - P'_k \oplus Q''_l(f)(x,y) = \int_a^b \int_c^d G_k(x,s) \cdot H_l(y,t) \cdot D^{k,l} f(s,t) \, ds \, dt$$

$$(\ f \in \mathscr{C}^{k,l}([a,b] \times [c,d]) \ ) \ .$$

In view of the properties of the remainder kernels $G_k(x,s)$ and $H_l(y,t)$ as listed in Section 5.4 the integral remainder formula (3) yields the *Cauchy remainder formula of bivariate polynomial blending interpolation* :

$$(4) \quad f(x,y) - P'_k \oplus Q''_l(f)(x,y) = (x-x_1) \cdots (x-x_k) \cdot (y-y_1) \cdots (y-y_l) \cdot \frac{D^{k,l} f(\xi,\eta)}{k! \cdot l!}$$

$$(\ \xi \in [a,b] \ , \ \eta \in [c,d] \ ) \ .$$

Recall that $b = a+h$ and $d = c+h$. It follows from (4) that the *order of convergence* of bivariate polynomial blending interpolation is $k+l$:

$$(5) \qquad f(x,y) - P'_k \oplus Q''_l(f)(x,y) = \mathcal{O}(h^{k+l}) \quad (\ h \to 0 \ ) \ .$$

Next we will determine the remainder of *bivariate polynomial tensor product interpolation* . Its interpolation projector is given by

$$P'_k Q''_l(f)(x,y) = \sum_{i=1}^{k} \sum_{j=1}^{l} f(x_i,y_j) \cdot \prod_{\substack{j=1 \\ j \neq i}}^{k} \frac{(x-x_j)}{(x_i-x_j)} \cdot \prod_{\substack{r=1 \\ r \neq j}}^{l} \frac{(y-y_r)}{(y_j-y_r)} \ .$$

Its remainder projector satisfies the equation

$$(P'_k Q''_l)^c = P'^{\,c}_k + Q''^{\,c}_l - P'^{\,c}_k \cdot Q''^{\,c}_l \ .$$

This implies the following *remainder formula of bivariate polynomial tensor product interpolation* :

(6)
$$f(x,y) - P'_k Q''_l(f)(x,y)$$
$$= \int_a^b G_k(x,s)\cdot D^{k,0}f(s,y)\ ds + \int_c^d H_l(y,t)\cdot D^{0,l}f(x,t)\ dt$$
$$- \int_a^b \int_c^d G_k(x,s)\cdot H_l(y,t)\cdot D^{k,l}f(s,t)\ ds\ dt$$
$$(\ f \in \mathcal{C}^{k,l}([a,b]\times[c,d])\ ) .$$

In view of the properties of the remainder kernels $G_k(x,s)$ and $H_l(y,t)$ as listed in Section 5.4 the integral remainder formula (6) yields the *Cauchy remainder formula of bivariate polynomial tensor product interpolation* :

(7)
$$f(x,y) - P'_k Q''_l(f)(x,y)$$
$$= (x-x_1)\cdots(x-x_k)\cdot\frac{D^{k,0}f(\xi,y)}{k!\cdot 0!} + (y-y_1)\cdots(y-y_l)\cdot\frac{D^{0,l}f(x,\eta)}{0!\cdot l!}$$
$$- (x-x_1)\cdots(x-x_k)\cdot(y-y_1)\cdots(y-y_l)\cdot\frac{D^{k,l}f(\sigma,\tau)}{k!\cdot l!}$$
$$(\ \xi,\sigma \in [a,b]\ ,\ \eta,\tau \in [c,d]\ ) .$$

Recall that $b = a+h$ and $d = c+h$. It follows from (7) that the *order of convergence* of bivariate polynomial tensor product interpolation is

$$q := \min\{k,l\} ,$$

i. e., we have

(8)
$$f(x,y) - P'_k Q''_l(f)(x,y) = \mathcal{O}(h^q) \quad (\ h \to 0\ ) .$$

5.6 The error in Biermann interpolation

We describe the remainder projectors of the univariate polynomial Lagrange interpolation projectors P_m and Q_n :

$$(1)\quad P_m^c(f)(x) = f(x)-P_m(f)(x) = \int_a^b G_{k_m}(x,s)D^{k_m}f(s)\,ds \quad (f \in \mathscr{C}^{k_m}([a,b])),$$

$$Q_n^c(g)(y) = g(y)-Q_n(g)(y) = \int_c^d H_{l_n}(y,t)D^{l_n}g(t)\,dt \quad (g \in \mathscr{C}^{l_n}([c,d])).$$

The kernel functions are given by

$$G_{k_m}(x,s) = M(x,x_1,\dots,x_{k_m},s)\cdot(x-x_1)\cdots(x-x_{k_m}) ,$$

$$H_{l_n}(y,t) = M(y,y_1,\dots,y_{l_n},t)\cdot(y-y_1)\cdots(y-y_{l_n}) .$$

Moreover we assume that the relations

$$b = a + h , \quad d = c + h \quad (h > 0),$$

$$a \le x_i \le b \quad (1 \le i \le k_r) , \quad c \le y_j \le d \quad (1 \le j \le l_r)$$

hold.

Recall that the Biermann interpolation projector B_r is defined by

$$B_r = P'_1Q''_r \oplus P'_2Q''_{r-1} \oplus \cdots \oplus P'_rQ''_1 .$$

The follwing Proposition 1 contains an integral representation of the remainder of Biermann interpolation.

Proposition 1. *If* $f \in \mathcal{C}^{k_r,l_r}([a,b]\times[c,d])$ *, then the identity*

$$(2) \quad f(x,y) - B_r(f)(x,y)$$

$$= \int_a^b G_{k_r}(x,s)\cdot D^{k_r,0} f(s,y)\, ds + \int_c^d H_{l_r}(y,t)\cdot D^{0,l_r} f(x,t)\, dt$$

$$+ \sum_{m=1}^{r-1} \int_a^b \int_c^d G_{k_{r-m}}(x,s)\cdot H_{l_m}(y,t)\cdot D^{k_{r-m},l_m} f(s,t)\, ds\, dt$$

$$- \sum_{m=1}^{r} \int_a^b \int_c^d G_{k_{r-m+1}}(x,s)\cdot H_{l_m}(y,t)\cdot D^{k_{r-m+1},l_m} f(s,t)\, ds\, dt \; .$$

holds.

Proof: It follows from Proposition 2 of Section 1.4 that the remainder of the Biermann interpolation projector is given by

$$(3) \qquad B_r^c = P'^c_r + P'^c_{r-1}\cdot Q''^c_1 + \ldots + P'^c_1\cdot Q''^r_{r-1} + Q''^c_r - (P'^c_r\cdot Q''^c_1 + \ldots + P'^c_1\cdot Q''^c_r).$$

Using the formulas (1) for the remainder projectors P^c_n and Q^c_n and the method of parametric extension we obtain (2) . This completes the proof of Proposition 1.

Reamark 1. For $r = 1$ the remainder formula (2) yields the remainder formula (6) in Section 5.5 of bivariate polynomial product interpolation .

In view of the properties of the remainder kernels $G_{k_m}(x,s)$ and $H_{l_n}(y,t)$ as listed in Section 5.4 the integral remainder formula (2) yields the *Cauchy remainder formula of bivariate polynomial Biermann interpolation* :

$$(4)\quad f(x,y) - B_r(f)(x,y)$$

$$= (x-x_1)\cdots(x-x_{k_r})\cdot\frac{D^{k_r,0}f(\xi_r,y)}{k_r!\,0!} + (y-y_1)\cdots(y-y_{l_r})\cdot\frac{D^{0,l_r}f(x,\eta_r)}{0!\,l_r!}$$

$$+ \sum_{m=1}^{r-1}\prod_{i=1}^{k_{r-m}}(x-x_i)\cdot\prod_{j=1}^{l_m}(y-y_j)\cdot\frac{D^{k_{r-m},l_m}f(\xi_{r-m},\eta_m)}{k_{r-m}!\,l_m!}$$

$$- \sum_{m=1}^{r}\prod_{i=1}^{k_{r+1-m}}(x-x_i)\cdot\prod_{j=1}^{l_m}(y-y_j)\cdot\frac{D^{k_{r+1-m},l_m}f(\sigma_{r+1-m},\tau_m)}{k_{r+1-m}!\,l_m!}$$

$$(\ \xi_i,\sigma_i \in [a,b]\ ,\ \eta_i,\tau_i \in [c,d]\quad (\ 1 \le i \le r\ )\ )\ .$$

Proposition 2. *Let*

$$(5)\qquad q = \min\{\ k_{r-m}+l_m\ :\ 0 \le m \le r\ \}$$

with $k_0 = 0$, $l_0 = 0$. *Then* q *is the order of convergence of bivariate polynomial Biermann interpolation :*

$$(6)\qquad f(x,y) - B_r(f)(x,y) = \mathcal{O}(h^q)\quad (\ h \to 0\ )\ .$$

Proof: Relation (6) is an easy consequence of relation (4) which is valid for smooth functions

$$f \in \mathcal{C}^{k_r,l_r}([a,b]\times[c,d])\ .$$

This completes the proof of Proposition 2.

Example 1. We assume

$$k_1 = l_1 = k \quad , \quad k_2 = l_2 = 2 \cdot k.$$

The special Biermann interpolation projector

$$B_2 \; = \; P'_1Q''_2 \oplus P'_2Q''_1$$

is called the projector of *discrete polynomial blending interpolation* . The discrete blending interpolation projector $P'_1Q''_2 \oplus P'_2Q''_1$ is obtained from the blending interpolation projector $P'_1 \oplus Q''_1$ by the approximation

$$P'_1 + Q''_1 - P'_1Q''_1 \;\simeq\; P'_1Q''_2 + P'_2Q''_1 - P'_1Q''_1 \; .$$

The order of convergence of the polynomial blending interpolation projector $P'_1 \oplus Q''_1$ is $2 \cdot k$:

(8) $$f(x,y) \; - \; P'_1 \oplus Q''_1(f)(x,y) \; = \; \mathcal{O}(h^{2 \cdot k}) \quad (h \to 0) \; .$$

The order of convergence of the polynomial product interpolation projector $P'_2Q''_2$ is also $2 \cdot k$:

(9) $$f(x,y) \; - \; P'_2Q''_2(f)(x,y) \; = \; \mathcal{O}(h^{2 \cdot k}) \quad (h \to 0) \; .$$

It follows from Proposition 2 that the order of convergence of the special Biermann interpolation projector $P'_1Q''_2 \oplus P'_2Q''_1$ is again $2 \cdot k$:

(10) $$f(x,y) \; - \; P'_1Q''_2 \oplus P'_2Q''_1(f)(x,y) \; = \; \mathcal{O}(h^{2 \cdot k}) \quad (h \to 0) \; .$$

Thus the interpolation projectors $P'_1Q''_2 \oplus P'_2Q''_1$ and $P'_2Q''_2$ possess the order of convergence $2 \cdot k$ of the blending projector $P'_1 \oplus Q''_1$. It follows from Proposition 3 of Section 5.1 that the following equations hold:

(11) $$\dim \mathcal{R}(P'_2Q''_2) \; = \; 4 \cdot k^2 \quad ,$$

(12) $$\dim \mathcal{R}(P'_1Q''_2 \oplus P'_2Q''_1) = 3\cdot k^2 .$$

In view of the relations (11) and (12) discrete blending interpolation via $P'_1Q''_2 \oplus P'_2Q''_1$ is more efficient than product interpolation via $P'_2Q''_2$.

Example 2. We determine the order of convergence of the classical Biermann interpolation projector B_r for triangular elements as discussed in Section 5.2 . It is characterized by the following choice :

$$x_i = \frac{i\cdot h}{r}, \quad y_j = \frac{j\cdot h}{r} \quad (1 \le i,j \le r , \quad h > 0),$$

$$k_m = m , \quad l_n = n \quad (1 \le m,n \le r).$$

It follows from Proposition 2 that the order of convergence of the Biermann interpolation projectors for triangular elements is r :

(13) $$f(x,y) - B_r(f)(x,y) = \mathcal{O}(h^r) \quad (h \to 0) .$$

Example 3. We determine the order of convergence of the Biermann interpolation projector B_r for the rectangular elements as discussed in Section 5.3 . In this case we have

$$x_{2i-1} = -\frac{(1+2(r-i))\cdot h}{(2r-1)\cdot 2}, \quad x_{2i} = -x_{2i-1},$$

$$y_{2j-1} = -\frac{(1+2(r-j))\cdot h}{(2r-1)\cdot 2}, \quad y_{2j} = -y_{2j-1}$$

$$(1 \le i,j \le r) \quad ; \quad k_m = 2m , \quad l_n = 2n \quad (1 \le m,n \le r).$$

Again it follows from Proposition 2 that the order of convergence of the Biermann interpolation projector for the rectangular elements is $2\cdot r$:

(14) $$f(x,y) - B_r(f)(x,y) = \mathcal{O}(h^{2\cdot r}) \quad (h \to 0) .$$

Notes, Section 5

Polynomial blending interpolation has its origin in computer aided geometric design (BIRKHOFF - GORDON [10] , GORDON [27]). Discretization of polynomial blending interpolants yields bivariate polynomial interpolants which occur in the finite element method (see GORDON - HALL [31], DELVOS - POSDORF [21]). Our treatment of Boolean bivariate polynomial interpolation yields a systematic approach to these interpolation schemes and its predecessors (see BIERMANN [8], DELVOS - POSDORF [22], STANCU [44]). The Boolean approach includes the explicit construction of fundamental functions and remainder formulas of special interpolation methods of the finite element method (see LANCASTER [35], LANCASTER - WATKINS [36], WATKINS [46,47]). In a number of papers we extended the method of Boolean bivariate polynomial Lagrange interpolation to bivariate polynomial Hermite interpolation and applied this method to interpolation schemes of the finite element method (see BASZENSKI - DELVOS - POSDORF [6], BASZENSKI - DELVOS - HACKENBERG [7], DELVOS -POSDORF [20], MELKES [38], WATKINS [46,47]). For multivariate extensions of Boolean interpolation we refer to GORDON [28,29] and DELVOS [12,13].

6 Boolean bivariate spline interpolation

6.1 Univariate spline interpolation

The bivariate polynomial interpolation schemes presented in Section 5 are usually applied as local approximation tools. In this section we use techniques of Boolean projector algebras to generate bivariate global approximation methods which are based on periodic spline interpolation . Here we briefly recall the basic properties of periodic splines which will be used in the sequel.

Let $J = [0,2\pi]$ and let $\mathcal{C}^k_{2\pi}(J)$ be the linear space of k-times continuously differentiable functions satisfying the *periodicity conditions* :

$$D^j f(0) = D^j f(2\pi) \quad (0 \le j \le k) .$$

A *periodic spline function* of degree $2p-1$, $p \in \mathbb{N}$, with joints at

$$x_j = \frac{2\pi j}{N} \quad (j \in \mathbb{Z})$$

is a function $\sigma \in \mathcal{C}^{2p-2}_{2\pi}(J)$ with the properties that σ is a polynomial of degree at most $2p-1$ in each subinterval

$$[x_j, x_{j+1}] \quad (j \in \mathbb{Z}) .$$

The space of periodic of degree $2p-1$ with joints at x_j , $j \in \mathbb{Z}$, will be denoted by $\mathcal{S}_N$. $\mathcal{S}_N$ is an N-dimensional linear subspace of $\mathcal{C}^{2p-2}_{2\pi}(J)$. The following result concerning periodic spline interpolation on a uniform mesh is well known .

Proposition 1. *If* $f \in \mathcal{C}_{2\pi}(J)$, *then there exists a unique spline function* $S_N(f) \in \mathcal{S}_N$ *with the property that*

$$S_N(f)(x_j) = f(x_j) \quad (0 \le j < N) .$$

The interpolating spline function possesses a *Lagrange representation*

(1) $$S_N(f) = \sum_{j=0}^{N-1} f(x_j)\, L_{j,N}(x)$$

where $L_{j,N} \in \mathcal{S}_N$ satisfies the duality relations

$$L_{j,N}(x_k) = \delta_{j,k} \quad (0 \le j,k < N) .$$

KORNEICUK [34] has derived an integral representation for the *remainder of periodic spline interpolation* :

$$f(x) - S_N(f)(x) = \frac{1}{2\pi} \int_0^{2\pi} G_N(x,s)\, D^{2p} f(s)\, ds \quad (f \in \mathcal{C}_{2\pi}^{2p}(J)) .$$

The kernel function $G_N(x,s)$ is given by

(2) $$G_N(x,s) = b_{2p}(x) - b_{2p}(x-s) - \sum_{j=0}^{N-1} L_{j,N}(x)\big(b_{2p}(x_j) - b_{2p}(x_j-s)\big)$$

where b_{2p} the Bernoulli mono-spline of degree $2p$. It possesses the Fourier expansion

$$b_{2p}(x) = \sum_{k \ne 0} (ik)^{-2p} e^{ikx} .$$

Moreover there exists a positive constant c with the property that

(3) $$\sup \{ \frac{1}{2\pi} \int_0^{2\pi} |G_N(x,s)|\, ds : 0 \le x \le 2\pi \} = \frac{c}{N^{2p}} .$$

As an easy consequence of the relations (3) and (2) we obtain

Proposition 2. *If* $f \in \mathcal{C}_{2\pi}^{2p}(J)$, *then the error in spline interpolation is*

(4) $$\| f - S_N(f) \|_\infty = \mathcal{O}(\frac{1}{N^{2p}}) \quad (N \to \infty) .$$

Remark 1. For simplicity we have only considered odd degree periodic splines on uniform meshes. Using the results of [17] the relations (3) and (4) can be extended to even degree periodic splines on uniform meshes if the corresponding interpolation problem is uniquely solvable. This is the case for odd N or for periodic midpoint splines of even degree.

6.2 Spline blending and tensor product interpolation.

We discuss the simplest methods of bivariate periodic spline interpolation such as blending and tensor product interpolation and derive the corresponding integral remainder formulas. The difference in approximation power is easily illustrated for these two interpolation schemes.

Let $\mathcal{C}^{k,k}_{2\pi}(J^2)$ be the linear space of bivariate continuous functions on J^2 such that the following properties are satisfied:

(i) $$D^{j,l}f \in \mathcal{C}(J^2) \qquad (0 \le j,l \le k) .$$

(ii) The partial derivatives $D^{j,l}f$ are periodic with respect to each variable, i. e.,

$$D^{j,l}f(0,y) = D^{j,l}f(2\pi,y) \qquad (0 \le y \le 2\pi) ,$$

$$D^{j,l}f(x,0) = D^{j,l}f(x,2\pi) \qquad (0 \le x \le 2\pi) ,$$

$$(0 \le j,l \le k) .$$

We consider the parametric extensions S'_N and S''_N of the periodic spline interpolation projector S_N . The projectors S'_N and S''_N commute on

$$\mathcal{C}_{2\pi}(J^2) := \mathcal{C}^{0,0}_{2\pi}(J^2) .$$

The projector of *bivariate spline tensor product interpolation* is

$$S'_N S''_N(f)(x,y) = \sum_{j=0}^{N-1} \sum_{k=0}^{N-1} f(jh,kh)\cdot L_j(x)\cdot L_k(y) \qquad (h = \frac{2\cdot\pi}{N})$$

with $L_j := L_{j,N}$. The bivariate periodic spline function $S'_N S''_N(f)$ satisfies the *interpolation conditions* :

$$S'_N S''_N(f)(jh,kh) = f(jh,kh) \qquad (0 \le j,k < N) .$$

The projector of *bivariate spline blending interpolation* is

$$S'_N \oplus S''_N(f)(x,y) = \sum_{j=0}^{N-1} f(jh,y)L_j(x) + \sum_{k=0}^{N-1} f(x,kh)L_k(y) - \sum_{j=0}^{N-1}\sum_{k=0}^{N-1} f(jh,kh)L_j(x)L_k(y).$$

The bivariate periodic spline blending function $S'_N \oplus S''_N(f)$ satisfies the *interpolation conditions* :

$$S'_N \oplus S''_N(f)(jh,y) = f(jh,y) \quad (0 \le j < N ,\ 0 \le y \le 2\pi) ,$$

$$S'_N \oplus S''_N(f)(x,kh) = f(x,kh) \quad (0 \le k < N ,\ 0 \le x \le 2\pi) .$$

We will determine the approximation power of the bivariate spline interpolation projectors $S'_N S''_N$ and $S'_N \oplus S''_N$.

Proposition 1. *If* $f \in \mathcal{C}^{2p,2p}_{2\pi}(J^2)$, *then the error in bivariate spline blending interpolation is*

(1) $$\|f - S'_N \oplus S''_N(f)\|_\infty = \mathcal{O}(\frac{1}{N^{4p}}) \quad (N \to \infty) .$$

Proof: Recall that

$$(S'_N \oplus S''_N)^c = S'^c_N S''^c_N .$$

Using relation (2) of Section 6.1 we obtain the integral representation of the *remainder of bivaraite spline blending interpolation* :

$$f(x,y) - S'_N \oplus S''_N(f)(x,y) = \frac{1}{4\pi^2} \int_0^{2\pi}\int_0^{2\pi} G_N(x,s)G_N(y,t)D^{2p,2p}f(s,t)\, ds\, dt .$$

Taking into account relation (3) of Section 6.1 we obatin (1) . This completes the proof of Proposition 1.

Proposition 2. *If* $f \in \mathcal{C}^{2p,2p}_{2\pi}(J^2)$, *then the error in bivariate spline tensor product interpolation is*

(2) $$\|f - S'_N S''_N(f)\|_\infty = \mathcal{O}(\frac{1}{N^{2p}}) \quad (N \to \infty) .$$

Proof: Recall that

$$(S'_N S''_N)^c = S'^c_N \oplus S''^c_N .$$

Using relation (2) of Section 6.1 we obtain the integral representation of the *remainder of bivariate spline tensor product interpolation*:

(3) $$f(x,y) - S'_N S''_N(f)(x,y)$$

$$= \frac{1}{2\pi}\int_0^{2\pi} G_N(x,s)D^{2p,0}f(s,y)\,ds + \frac{1}{2\pi}\int_0^{2\pi} G_N(x,t)D^{0,2p}f(x,t)\,dt$$

$$- \frac{1}{4\pi^2}\int_0^{2\pi}\int_0^{2\pi} G_N(x,s)\cdot G_N(y,t)\cdot D^{2p,2p}f(s,t)\,ds\,dt .$$

Again relation (3) of Section 6.1 yields (3) which completes the proof of Proposition 2.

6.3 r-th order spline blending interpolation

Let N be a positive integer which is greater than 1 . The univariate spline projectors

$$S_{N^m} \quad (m \in \mathbb{N})$$

commute on $\mathscr{C}_{2\pi}(J)$ and satisfy the order relation

(1) $$S_{N^m} \leq S_{N^n} \quad (m \leq n) .$$

The set of interpolation points of S_{N^m} is

$$\mathcal{P}(S_{N^m}) = \{ jh_m : 0 \leq j < N^m \}$$

with

$$h_m = \frac{2\pi}{N^m} .$$

The parametric extensions

$$S'_{N^m} , \quad S''_{N^n} \quad (m,n \in \mathbb{N})$$

generate a distributive lattice of interpolation projectors on $\mathscr{C}_{2\pi}(J^2)$ and a related Boolean algebra of projectors containing the remainder projectors. The projector of *bivariate* r*-th order spline blending interpolation is*

$$\mathbf{B}_r^N = S'_{N^1} \cdot S''_{N^r} \oplus S'_{N^2} \cdot S''_{N^{r-1}} \oplus \cdots \oplus S'_{N^r} \cdot S''_{N^1} .$$

Proposition 1. *The set of interpolation points of the projector of* r*-th order spline blending interpolation is*

(2) $$\mathcal{P}(\mathbf{B}_r^N) = \bigcup_{m=1}^{r} \{ (jh_m, kh_{r+1-m}) : 0 \leq j < N^m, \ 0 \leq k < N^{r+1-m} \} .$$

Proof: We have

$$\mathcal{P}(S'^m_N \cdot S''^n_N) = \{ (jh_m, kh_n) : 0 \le j < N^m,\ 0 \le k < N^n \}.$$

Using Proposition 2 of Section 2.3 we obtain

$$\mathcal{P}(\mathbf{B}^N_r) = \bigcup_{m=1}^{r} \mathcal{P}(S'^m_N \cdot S''^{r+1-m}_N)$$

which completes the proof of Propostion 1 .

To determine the *approximation power* of the interpolation projector $\mathbf{B}^N_r$ of r–th order spline blending interpolation we need the following

Proposition 2. *If* $f \in \mathscr{C}^{2p,2p}_{2\pi}(J^2)$, *then the remainder of* r*–th order spline blending interpolation is*

(3) $f(x,y) - \mathbf{B}^N_r(f)(x,y)$

$$= \frac{1}{2\pi}\int_0^{2\pi} G_{N^r}(x,s)\cdot D^{2p,0}f(s,y)\,ds + \frac{1}{2\pi}\int_0^{2\pi} G_{N^r}(x,t)\ D^{0,2p}f(x,t)\,dt$$

$$+ \sum_{m=1}^{r-1} \frac{1}{4\pi^2}\int_0^{2\pi}\int_0^{2\pi} G_{N^{r-m}}(x,s)\cdot G_{N^m}(y,t)\cdot D^{2p,2p}f(s,t)\,ds\,dt$$

$$- \sum_{m=1}^{r} \frac{1}{4\pi^2}\int_0^{2\pi}\int_0^{2\pi} G_{N^{r+1-m}}(x,s)\cdot G_{N^m}(y,t)\cdot D^{2p,2p}f(s,t)\,ds\,dt .$$

Proof: It follows from (1) that

$$S'^m_N \le S'^n_N , \quad S''^m_N \le S''^n_N \qquad (m \le n) .$$

Thus, we can apply Proposition 2 of Section 1.4 and obtain

$$(4)\qquad I-B^N_r = (S'_Nr)^c + \sum_{m=1}^{r-1}(S'_Nr-m)^c(S''_Nm)^c + (S''_Nr)^c - \sum_{m=1}^{r}(S'_Nr{+}1{-}m)^c(S''_Nm)^c.$$

Since

$$(S'_Nr)^c(f)(x,y) = \frac{1}{2\pi}\int_0^{2\pi} G_{N^r}(x,s)\cdot D^{2p,0}f(s,y)\, ds \quad ,$$

$$(S''_Nr)^c(f)(x,y) = \frac{1}{2\pi}\int_0^{2\pi} G_{N^r}(x,t)\; D^{0,2p}f(x,t)\, dt \quad ,$$

and

$$(S'_Nm)^c(S''_Nn)^c(f)(x,y) = \frac{1}{4\pi^2}\int_0^{2\pi}\int_0^{2\pi} G_{N^m}(x,s)\cdot G_{N^n}(y,t)\cdot D^{2p,2p}f(s,t)\, ds\, dt \quad ,$$

formula (3) follows from relation (4). This completes the proof of Proposition 2.

Remark 1. Choosing r = 1 in Proposition 2 we obtain the remainder formula of bivariate spline product interpolation.

We consider the case r = 2 . The projector $\mathbf{B}^N_2$ is given by

$$(5)\qquad \mathbf{B}^N_2 = S'_N1\cdot S''_N2 + S'_N2\cdot S''_N1 - S'_N1\cdot S''_N1 \ .$$

$\mathbf{B}^N_2$ is called the projector of *discrete spline blending interpolation.* $\mathbf{B}^N_2$ is obtained from the projector $S'_N\oplus S''_N$ of spline blending interpolation by the approximation

$$S'_N + S''_N - S'_N\cdot S''_N \simeq S'_N1\cdot S''_N2 + S'_N2\cdot S''_N1 - S'_N1\cdot S''_N1 \ .$$

Proposition 3. *If* $f \in \mathcal{C}^{2p,2p}_{2\pi}(J^2)$, *then the error in bivariate discrete spline blending interpolation is*

(6) $$\|f - \mathbf{B}^N_2(f)\|_\infty = \mathcal{O}(\frac{1}{N^{4p}}) \qquad (N \to \infty) .$$

Proof: We apply relation (3) of Section 6.2 to (3) and obtain

$$|f(x,y) - \mathbf{B}^N_2(f)(x,y)|$$

$$\leq \frac{1}{2\pi}\int_0^{2\pi} |G_{N^2}(x,s)|\,|D^{2p,0}f(s,y)|\ ds + \frac{1}{2\pi}\int_0^{2\pi} |G_{N^2}(x,t)|\,|D^{0,2p}f(x,t)|\ dt$$

$$+ \frac{1}{4\pi^2}\int_0^{2\pi}\int_0^{2\pi} |G_{N^1}(x,s)|\cdot|G_{N^1}(y,t)|\cdot|D^{2p,2p}f(s,t)|\ ds\ dt$$

$$+ \frac{1}{4\pi^2}\int_0^{2\pi}\int_0^{2\pi} |G_{N^2}(x,s)|\cdot|G_{N^1}(y,t)|\cdot|D^{2p,2p}f(s,t)|\ ds\ dt$$

$$+ \frac{1}{4\pi^2}\int_0^{2\pi}\int_0^{2\pi} |G_{N^1}(x,s)|\cdot|G_{N^2}(y,t)|\cdot|D^{2p,2p}f(s,t)|\ ds\ dt$$

$$\leq \frac{1}{2\pi}\int_0^{2\pi} |G_{N^2}(x,s)|\cdot ds\ \ \|D^{2p,0}f\|_\infty + \frac{1}{2\pi}\int_0^{2\pi} |G_{N^2}(x,t)|\ dt\ \ \|D^{0,2p}f\|_\infty$$

$$+ \frac{1}{4\pi^2}\int_0^{2\pi}\int_0^{2\pi} |G_{N^1}(x,s)|\cdot|G_{N^1}(y,t)|\ ds\ dt\ \ \|D^{2p,2p}f\|_\infty$$

$$+ \frac{1}{4\pi^2}\int_0^{2\pi}\int_0^{2\pi} |G_{N^2}(x,s)|\cdot|G_{N^1}(y,t)|\ ds\ dt\ \ \|D^{2p,2p}f\|_\infty$$

$$+ \frac{1}{4\pi^2}\int_0^{2\pi}\int_0^{2\pi} |G_{N^1}(x,s)|\cdot|G_{N^2}(y,t)|\ ds\ dt\ \ \|D^{2p,2p}f\|_\infty$$

$$\leq \frac{c}{N^{4p}} \cdot \|D^{2p,0}f\|_\infty + \frac{c}{N^{4p}} \cdot \|D^{0,2p}f\|_\infty$$

$$+ \frac{c}{N^{2p}} \cdot \frac{c}{N^{2p}} \cdot \|D^{2p,2p}f\|_\infty + 2 \cdot \frac{c}{N^{4p}} \cdot \frac{c}{N^{2p}} \cdot \|D^{2p,2p}f\|_\infty$$

$$= \mathcal{O}\left(\frac{1}{N^{4p}}\right) .$$

This completes the proof of Proposition 3.

It follows Propositions 1 and 2 of Section 6.2 and Proposition 3 that the bivariate spline interpolation projectors

$$S'_N \oplus S''_N \quad , \quad S'_N 1 \cdot S''_N 2 \oplus S'_N 2 \cdot S''_N 1 \quad , \quad S'_N 2 \cdot S''_N 2$$

possess the same order of approximation $\mathcal{O}(\frac{1}{N^{4p}})$. Since

(7)
$$\dim \mathcal{R}(S'_N 1 \cdot S''_N 2 \oplus S'_N 2 \cdot S''_N 1) = 2 \cdot N^3 - N^2,$$
$$\dim \mathcal{R}(S'_N 2 \cdot S''_N 2) = N^4 ,$$

it follows that the discrete spline blending interpolation projector $S'_N 1 \cdot S''_N 2 \oplus S'_N 2 \cdot S''_N 1$ is more efficient than the spline product interpolation projector $S'_N 2 \cdot S''_N 2$.

In the following we will determine the approximation power of the r-*th spline blending interpolation projector* $\mathbf{B}^2_r$.

Proposition 3. *If* $f \in \mathcal{C}^{2p,2p}_{2\pi}(J^2)$, *then the error in* r-*th order spline blending interpolation* $\mathbf{B}^2_r$ *is*

(8)
$$\|f - \mathbf{B}^2_r(f)\|_\infty = \mathcal{O}\left(\frac{r}{(2^r)^{2p}}\right) \quad (r \to \infty) .$$

Proof: We apply relation (3) of Section 6.2 to (3) and obtain

$$|f(x,y) - \mathbf{B}_r^2(f)(x,y)|$$

$$\le \frac{1}{2\pi}\int_0^{2\pi} |G_{2^r}(x,s)|\,|D^{2p,0}f(s,y)|\;ds + \frac{1}{2\pi}\int_0^{2\pi} |G_{2^r}(x,t)|\,|D^{0,2p}f(x,t)|\;dt$$

$$+ \sum_{m=1}^{r-1} \frac{1}{4\pi^2}\int_0^{2\pi}\int_0^{2\pi} |G_{2^{r-m}}(x,s)|\,|G_{2^m}(y,t)|\,|D^{2p,2p}f(s,t)|\;ds\;dt$$

$$+ \sum_{m=1}^{r} \frac{1}{4\pi^2}\int_0^{2\pi}\int_0^{2\pi} |G_{2^{r+1-m}}(x,s)|\,|G_{2^m}(y,t)|\,|D^{2p,2p}f(s,t)|\;ds\;dt$$

$$\le \frac{1}{2\pi}\int_0^{2\pi} |G_{2^r}(x,s)|\;ds\cdot\|D^{2p,0}f\|_\infty + \frac{1}{2\pi}\int_0^{2\pi} |G_{2^r}(x,t)|\;dt\cdot\|D^{0,2p}f\|_\infty$$

$$+ \sum_{m=1}^{r-1} \frac{1}{4\pi^2}\int_0^{2\pi}\int_0^{2\pi} |G_{2^{r-m}}(x,s)|\,|G_{2^m}(y,t)|\;ds\;dt\cdot\|D^{2p,2p}f\|_\infty$$

$$+ \sum_{m=1}^{r} \frac{1}{4\pi^2}\int_0^{2\pi}\int_0^{2\pi} |G_{2^{r+1-m}}(x,s)|\,|G_{2^m}(y,t)|\;ds\;dt\cdot\|D^{2p,2p}f\|_\infty$$

$$\le \frac{c}{(2^r)^{2p}}\cdot\|D^{2p,0}f\|_\infty + \frac{c}{(2^r)^{2p}}\cdot\|D^{0,2p}f\|_\infty$$

$$+ \sum_{m=1}^{r-1} \frac{c}{(2^{r-m})^{2p}}\cdot\frac{c}{(2^m)^{2p}}\cdot\|D^{2p,2p}f\|_\infty$$

$$+ \sum_{m=1}^{r} \frac{c}{(2^{r-m+1})^{2p}}\cdot\frac{c}{(2^m)^{2p}}\cdot\|D^{2p,2p}f\|_\infty$$

$$= \mathcal{O}\left(\frac{r}{(2^r)^{2p}}\right) .$$

This completes the proof of Proposition 4.

We apply Proposition 3 of Section 5.1 to determine the dimension of $\mathcal{R}(\mathbf{B}_r^2)$. We have

$$\dim \mathcal{R}(\mathbf{B}_r^2)$$

$$= \sum_{m=0}^{r-1} \dim \mathcal{R}(S'_{2^m} \cdot S''_{2^{r+1-m}} - S'_{2^m} \cdot S''_{2^{r-m}}) + \dim \mathcal{R}(S'_{2^r} \cdot S''_{2^1})$$

$$= \sum_{m=0}^{r-1} (2^m 2^{r+1-m} - 2^m 2^{r-m}) + 2^{r+1}$$

$$= (r-1) \cdot 2^r + 2^r \cdot 2 ,$$

i. e.,

$$\text{(9)} \qquad \dim \mathcal{R}(\mathbf{B}_r^2) = (r+1) \cdot 2^r .$$

Since

$$\text{(10)} \qquad \dim \mathcal{R}(S'_{2^r} \cdot S''_{2^r}) = (2^r)^2$$

and

$$\text{(11)} \qquad \| f - S'_{2^r} S''_{2^r}(f) \|_\infty = \mathcal{O}\left(\frac{1}{(2^r)^{2p}}\right) \qquad (r \to \infty),$$

the projector $\mathbf{B}_r^2$ of r-th order spline blending interpolation is more efficient than the projector $S'_{2^r} \cdot S''_{2^r}$ of bivariate spline product interpolation .

Notes, Section 6

Spline - blended bivariate interpolation was first considered by GORDON [27] . Discrete spline - blended bivariate interpolants were introduced by GORDON [28]. Our representation follows BASZENSKI - DELVOS [5] (see also BASZENSKI [2] and DELVOS - POSDORF [19]). We note that bivariate spline interpolation has been treated in the functional analytic framework of Hilbert spaces via the theory of generalized inverses ([11, 15, 16,18]).

7 Boolean Fourier approximation

7.1 The univariate Korobov space

The theory of right invertible operators provides an abstract formulation of the classical Taylor formula based on definite integration and point evaluation. In the following we will present a more sophisticated application of the abstract Taylor formula based on convolution and differences. The univariate Korobov space is shown to be related to the remainder term.

Let $L^2(J)$ denote the Hilbert space of square integrable functions defined on the interval $J = [0,2\pi]$. The inner product of $L^2(J)$ is given by

$$(f,g) = \frac{1}{2\pi}\int_0^{2\pi} f(x)\cdot\overline{g(x)}\,dx \quad .$$

The exponential functions

$$e_k(x) = e^{ikx} \qquad (k \in \mathbb{Z})$$

form an orthonormal basis of $L^2(J)$. It is well known that every smooth function $f \in \mathscr{C}^m(J)$ with $m \in \mathbb{N}$ possesses a convergent Fourier series :

$$f(x) = \sum_{m=-\infty}^{\infty} (f,e_m)e_m(x) \quad (0 < x < 2\pi) \ , \qquad \frac{f(0) + f(2\pi)}{2} = \sum_{m=-\infty}^{\infty} (f,e_m) \ .$$

The *Bernoulli functions* b_m are defined recursively :

(1) $$b_1(x) = \pi - x \ , \quad Db_m(x) = b_{m-1}(x) \ , \quad (b_m,e_0) = 0$$

$$(0 < x < 2\pi \ , \ m \in \mathbb{N}).$$

The Fourier series of b_m is given by

(2) $$b_m(x) = \sum_{\substack{k=-\infty \\ k\neq 0}}^{\infty} (ik)^{-m}\, e_k(x) \quad .$$

The *convolution* of $f,g \in L^2(J)$ is defined by

$$f*g(x) \;=\; \frac{1}{2\pi}\int_0^x f(x-y)\cdot g(y)\,dy \;+\; \frac{1}{2\pi}\int_x^{2\pi} f(x-y+2\pi)\cdot g(y)\,dy \qquad (\,0 \le x \le 2\pi\,).$$

The following relations hold:

(3) $$f*g \;\in\; \mathcal{C}_{2\pi}(J) \;,$$

(4) $$(f*g,e_k) \;=\; (f,e_k)\cdot(g,e_k) \qquad (\,k \in \mathbb{Z}\,) \;.$$

This implies

(5) $$f*g \;\in\; A(J)$$

where $A(J)$ denotes the *Wiener algebra* of functions $f \in \mathcal{C}_{2\pi}(J)$ which possess an absolutely convergent Fourier series :

$$\sum_{m=-\infty}^{\infty} |(f,e_m)| \;<\; \infty \;.$$

We use the Bernoulli function b_1 and the operation of convolution to define the operator R on $\mathcal{C}(J)$:

$$R(f) \;=\; b_1*f + (f,e_0) \;.$$

It useful to introduce the complex numbers

$$\lambda_k \;=\; \frac{1}{i\cdot k} \qquad (\,k \in \mathbb{Z}\;,\; k \ne 0\,) \;,\qquad \lambda_0 \;=\; 1 \;.$$

The operator R possesses the eigenfunctions e_k , $k \in \mathbb{Z}$, with eigenvalues λ_k:

(6) $$R(e_k) \;=\; \lambda_k\cdot e_k \qquad (\,k \in \mathbb{Z}\,) \;.$$

Thus, R is an injective operator on $\mathcal{C}(J)$ with the property that

(7) $$\mathcal{R}(R) = \mathcal{C}^1(J) \cap \mathcal{C}_{2\pi}(J) \quad .$$

Next we introduce the operator P on $\mathcal{C}(J)$ by

$$P(f) = \frac{f(0) - f(2\pi)}{2\pi} \cdot b_1 \quad .$$

Then we have

(8) $$P(f) = -(Df,e_0)b_1 \qquad (f \in \mathcal{C}^1(J)) \quad .$$

It is easily seen that P is a projector on $\mathcal{C}(J)$ which possesses the property

(9) $$PR = 0 \quad .$$

It follows from Proposition 1 of Section 4.1 that the operators R and P define a unique right invertible operator A with right inverse R and initial projector P .

Proposition 1. *The right invertible operator* A *satisfies*

(10) $$A(f) = Df + (f-Df,e_0)$$

with $f \in \mathcal{D}(A) = \mathcal{C}^1(J)$.

Proof: We have

$$R(Df + (f-Df,e_0)) = R(Df) + (f-Df,e_0) = b_1{*}Df + (Df,e_0) + (f-Df,e_0) \, ,$$

i. e., we have

(11) $$(R(Df + (f-Df,e_0)) = b_1{*}Df + (f,e_0) \quad .$$

Next we can conclude

$$b_1 * Df(x)$$

$$= \frac{1}{2\pi} \int_0^x b_1(x-y) \cdot Df(y)\, dy + \frac{1}{2\pi} \int_x^{2\pi} b_1(x-y+2\pi) \cdot Df(y)\, dy$$

$$= \frac{1}{2\pi} \left(b_1(0) \cdot f(x) - b_1(x) \cdot f(0) - \int_0^x f(y)\, dy \right)$$

$$+ \frac{1}{2\pi} \left(b_1(x) \cdot f(2\pi) - b_1(2\pi) \cdot f(x) - \int_x^{2\pi} f(y)\, dy \right)$$

$$= f(x) - \frac{f(0) - f(2\pi)}{2\pi} \cdot b_1(x) - (f,e_0) \quad .$$

This implies

$$(12) \qquad b_1 * Df(x) + (f,e_0) = f(x) - P(f)(x) \quad .$$

Since $A(f)$ is the unique solution of the equation

$$R(g) = f - P(f) \quad ,$$

we obtain

$$g = A(f) = Df + (f-Df,e_0)$$

which completes the proof of Proposition 1.

We assume that a is a positive real number with the property that $a \geq 1$. We use the eigenvalues of the operator R to define the univariate *Korobov space* $E^a(J)$:

$$E^a(J) = \{ f \in L^2(J) : (f,e_k) = \mathcal{O}(|\lambda_k|^a) \quad (|k| \to \infty) \} \quad .$$

Proposition 2. *The inclusion*

(13) $$\mathscr{C}^1(J) \subseteq E^1(J)$$

is true.

Proof: It follows from (12) that the relation

$$f = b_1 * Df(x) + (f,e_0) - (Df,e_0)b_1$$

holds. It follows from (2) that $b_1 \in E^1(J)$. Taking into account relation (4) we obtain

$$(b_1 * Df, e_k) = (b_1, e_k) \cdot (Df, e_k) \quad (k \in \mathbb{Z})$$

which implies $f \in E^1(J)$. This completes the proof of Proposition 1.

Proposition 3. *If* $f \in E^a(J)$ *and* $g \in E^a(J)$, *then*

(14) $$f * g \in E^{a+b}(J) \quad .$$

Proof: The assumptions imply

$$(f,e_k) = O(|\lambda_k|^a) \quad , \quad (g,e_k) = O(|\lambda_k|^b) \quad (k| \to \infty).$$

Thus we obtain

$$(f,e_k) \cdot (g,e_k) = O(|\lambda_k|^{a+b}) \quad (|k| \to \infty)$$

whence relation (14) follows. This completes the proof of Proposition 3.

Proposition 4. *The inclusion*

$$\mathcal{C}^2(J) \cap \mathcal{C}_{2\pi}(J) \subseteq E^2(J) \tag{15}$$

is true.

Proof: It follows from relation (12) and

$$f \in \mathcal{C}^2(J) \cap \mathcal{C}_{2\pi}(J)$$

that

$$f(x) = b_1{*}Df(x) + (f,e_0) \quad .$$

It follows from Proposition 2 that

$$Df \in E^1(J) \quad .$$

Since $b_1 \in E^1(J)$, it follows from Proposition 3 that $f \in E^2(J)$. This completes the proof of Proposition 4.

Next we will apply Proposition 5 of Section 4.2 to derive an extension on Proposition 4. Recall that

$$f = P_m(f) + R^m A^m(f) \qquad (f \in \mathcal{D}(A^m)) \tag{16}$$

with

$$P_m = \sum_{j=0}^{m-1} R^j P A^j \ . \tag{17}$$

P_m is the initial projector of the right invertible operator A^m which possesses R^m as right inverse.

Proposition 5. *If* $f \in \mathcal{D}(A^m)$ *, then*

$$A^m(f) = D^m f + (f - D^m f, e_0) \quad . \tag{18}$$

Proof: We proceed by induction :

$$A^m(f)$$

$$= A(A^{m-1}f)$$

$$= DA^{m-1}(f) + (A^{m-1}f - DA^{m-1}(f),e_0)$$

$$= D^m f + (D^{m-1}f + (f-D^{m-1}f,e_0) - D^m f,e_0)$$

$$= D^m f + (D^{m-1}f,e_0) + (f-D^{m-1}f,e_0) - (D^m f,e_0)$$

$$= D^m f + (f-D^m f,e_0) \quad .$$

This completes the proof of Proposition 5 .

Proposition 6. *If* $f \in \mathcal{C}(J)$ *, then*

(19) $$R^m(f) = b_m{*}f + (f,e_0) \quad .$$

Proof: Again we proceed by induction . Recall that

$$b_m = b_{m-1}{*}b_1 = b_1{*}b_{m-1} \quad , \quad (b_m,e_0) = 0.$$

Then we can conclude

$$R^m(f)$$

$$= R(R^{m-1}f)$$

$$= b_1{*}R^{m-1}(f) + (R^{m-1}f,e_0)$$

$$= b_1{*}(b_{m-1}{*}f) + b_1{*}(f,e_0) + (b_{m-1}{*}f + (f,e_0),e_0)$$

$$= (b_1{*}b_{m-1}){*}f + (f,e_0)$$

$$= b_m * f + (f, e_0)$$

which completes the proof of Proposition 6.

Proposition 7. *If* $f \in \mathscr{C}^m(J)$, *then the relation*

$$f = \sum_{j=0}^{m-1} \frac{D^j f(0) - D^j f(2\pi)}{2\pi} \cdot b_{j+1} + b_m * D^m f + (f, e_0) \tag{20}$$

holds.

Proof: Using (18) and (19) we obtain

$$R^m A^m(f)$$

$$= b_m * A^m(f) + (A^m(f), e_0)$$

$$= b_m * D^m f + b_m * (f - D^m f, e_0) + (D^m f, e_0) + ((f - D^m f, e_0), e_0)$$

$$= b_m * D^m f + (f, e_0) ,$$

i. e. , we have

$$R^m A^m(f) = b_m * D^m f + (f, e_0) \quad (f \in \mathscr{C}^m(J)) . \tag{21}$$

Moreover, we can conclude

$$P_m(f)$$

$$= \sum_{j=0}^{m-1} R^j P A^j(f)$$

$$= \sum_{j=0}^{m-1} R^j P(D^j f + (f - D^j f, e_0))$$

$$= \sum_{j=0}^{m-1} \frac{D^j f(0) - D^j f(2\pi)}{2\pi} \cdot R^j(b_1)$$

$$= \sum_{j=0}^{m-1} \frac{D^j f(0) - D^j f(2\pi)}{2\pi} \cdot b_{j+1} \quad ,$$

i. e., we have

(22) $$P_m(f) = \sum_{j=0}^{m-1} \frac{D^j f(0) - D^j f(2\pi)}{2\pi} \cdot b_{j+1} \quad .$$

This completes the proof of Proposition 6.

Remark 1. It follows from relation (22) that

(23) $$\mathcal{N}(P_m) = \mathcal{C}^{m-1}_{2\pi}(J) \quad .$$

Proposition 8. *The inclusion*

(24) $$\mathcal{C}^{m+1}(J) \cap \mathcal{C}^{m-1}_{2\pi}(J) \subseteq E^{m+1}(J)$$

is true for every $m \in \mathbb{N}$.

Proof: It follows from (23) and (20) that the equality

$$f = b_m * D^m f + (f, e_0)$$

holds. Since $b_m \in E^m(J)$ and $D^m f \in E^1(J)$, an application of Proposition 3 yields (24) . This completes the proof of Proposition 8 .

Remark 2. Proposition 4 is obtained from Proposition 8 with $m = 1$.

7.2 Univariate Fourier approximation

The approximation properties of the Fourier partial sum projector are investigated in Korobov spaces. The periodic Taylor formula of Section 7.1 is used to construct the Krylov–Lanczos approximant of a smooth nonperiodic function.

Let $m \in \mathbb{Z}_+$. Then F_m denotes the *Fourier partial sum projector* which is defined by

$$F_m(f) = \sum_{k=-m}^{m} (f,e_k) \cdot e_k$$

where $f \in L^2(J)$. It is easily seen that F_m is a projector on the Korobov space $E^a(J)$.

Proposition 1. *If* $f \in E^a(J)$, *then*

(1) $$\|f - F_m(f)\| = \mathcal{O}(m^{-a+1/2}) \qquad (m \to \infty) .$$

Proof: It follows from Parseval's relation that

$$\|f - F(f)\|^2 = \sum_{|k|>m} |(f,e)|^2 .$$

Since $f \in E^a(J)$, we have

$$|(f,e_k)|^2 = \mathcal{O}(|\lambda_k|^{2a}) \qquad (|k| \to \infty) .$$

Then we can conclude

$$\|f - F_m(f)\|^2 = \mathcal{O}\left(\sum_{k=m+1}^{\infty} k^{-2a} \right) = \mathcal{O}(m^{-2a+1})$$

as $m \to \infty$. This completes the proof of Proposition 1.

There is a similar result in the uniform norm.

Proposition 2. *If* $f \in E^a(J)$ *with* $a > 1$, *then*

(2) $$\|f - F_m(f)\|_\infty = \mathcal{O}(m^{-a+1}) \quad (m \to \infty) .$$

Proof: Since $a > 1$ we have $E^a(J) \subseteq A(J)$. Then we obtain

$$\|f - F_m(f)\|_\infty = \mathcal{O}\left(\sum_{|k|>m} |(f,e)| \right) = \mathcal{O}\left(\sum_{k>m}^{\infty} k^{-a} \right) = \mathcal{O}(m^{-a+1}) \quad (m \to \infty)$$

which completes the proof of Proposition 2.

Next we discuss some special cases of Proposition 1 and Proposition 2 .

Proposition 3. *If* $f \in \mathcal{C}^1(J)$, *then*

(3) $$\|f - F_m(f)\| = \mathcal{O}(m^{-1/2}) \quad (m \to \infty) .$$

Proof: It follows from Proposition 2 of Section 7.1 that $f \in E^1(J)$. An application of Proposition 1 with $a = 1$ yields relation (3) . This completes the proof of Proposition 3.

Proposition 4. *If* $f \in \mathcal{C}^{q+1}(J) \cap \mathcal{C}^{q-1}_{2\pi}(J)$ *with* $q \in \mathbb{N}$, *then*

(4) $$\|f - F_m(f)\| = \mathcal{O}(m^{-q-1/2}) \quad (m \to \infty) .$$

Proof: It follows from Proposition 8 of Section 7.1 that $f \in E^{q+1}(J)$. An application of Proposition 1 with $a = q+1$ yields relation (4) . This completes the proof of Proposition 4 .

Proposition 5. *If* $f \in \mathscr{C}^{q+1}(J) \cap \mathscr{C}_{2\pi}^{q-1}(J)$ *with* $q \in \mathbb{N}$ *, then*

(5) $$\|f - F_m(f)\|_\infty = \mathcal{O}(m^{-q}) \quad (m \to \infty) .$$

Proof: It follows from Proposition 8 of Section 7.1 that $f \in E^{q+1}(J)$. Then we obtain

$$\|f - F_m(f)\|_\infty = \mathcal{O}\left(\sum_{|k|>m} |(f,e)| \right) = \mathcal{O}\left(\sum_{k>m}^{\infty} k^{-q-1} \right) = \mathcal{O}(m^{-q}) \quad (m \to \infty)$$

which completes the proof of Proposition 5 .

For $f \in \mathscr{C}^{q+1}(J)$ we introduce the *Krylov-Lanczos approximant* of f which is defined by

$$K_m^q(f) := P_q(f) + F_m(f - P_q(f))$$

with

$$P_q(f) = \sum_{j=0}^{q-1} \frac{D^j f(0) - D^j f(2\pi)}{2\pi} \cdot b_{j+1} .$$

The approximation properties of the Krylov-Lanczos approximant are described in the following propositions.

Proposition 6. *If* $f \in \mathscr{C}^{q+1}(J)$ *with* $q \in \mathbb{Z}_+$ *, then*

(6) $$\|f - K_m^q(f)\| = \mathcal{O}(m^{-q-1/2}) \quad (m \to \infty) .$$

Proof: We have

$$f - P_q(f) \in \mathscr{C}^{q+1}(J) \cap \mathscr{C}_{2\pi}^{q-1}(J)$$

and

$$f - K_m^q(f) = f - P_q(f) - F_m(f - P_q(f)) \quad .$$

An application of Proposition 4 yields the error relation (6) which completes the proof of Proposition 6 .

Proposition 7. *If* $f \in \mathscr{C}^{q+1}(J)$ *with* $q \in \mathbb{N}$ *, then*

(7) $$\|f - K_m^q(f)\|_\infty = \mathcal{O}(m^{-q}) \qquad (m \to \infty) \quad .$$

Proof: Again we have

$$f - P_q(f) \in \mathscr{C}^{q+1}(J) \cap \mathscr{C}_{2\pi}^{q-1}(J)$$

and

$$f - K_m^q(f) = f - P_q(f) - F_m(f - P_q(f)).$$

An application of Proposition 5 yields the error relation (7) which completes the proof of Proposition 6 .

7.3 The bivariate Korobov space

The methods presented in Section 4 are applied to develop a periodic Taylor formula for bivariate functions. The corresponding remainder term leads naturally to Korobov spaces of bivariate periodic functions.

We denote by $L^2(J^2)$ the Hilbert space of square integrable functions on J^2. The functions

$$e_{k,l} = e_k \otimes e_l \qquad (k,l \in \mathbb{Z})$$

form an orthonormal basis with respect to the inner product

$$(f,g) = \frac{1}{(2\pi)^2} \int_0^{2\pi} \int_0^{2\pi} f(x,y) \cdot \overline{g(x,y)}\, dx\, dy \qquad .$$

If $f,g \in L^2(J)$, then

$$f \otimes g \in L^2(J^2)$$

and we have

$$(1) \qquad (f \otimes g, e_{k,l}) = (f,e_k) \cdot (g,e_l) \qquad (k,l \in \mathbb{Z}) \quad .$$

The *convolution* of $f,g \in L^2(J^2)$ is defined as in the univariate case :

$$f*g(x,y)$$

$$= \frac{1}{(2\pi)^2} \left(\int_0^x \int_0^y f(x-s,y-t)g(s,t)dsdt + \int_x^{2\pi} \int_0^y f(x-s+2\pi,y-t)g(s,t)dsdt \right.$$

$$\left. + \int_0^x \int_y^{2\pi} f(x-s,y-t+2\pi)g(s,t)dsdt + \int_x^{2\pi} \int_y^{2\pi} f(x-s+2\pi,y-t+2\pi)g(s,t)dsdt \right) .$$

Then we have

$$(2) \qquad (f*g,e_{k,l}) = (f,e_{k,l})\cdot(g,e_{k,l}) \qquad (k,l \in \mathbb{Z}) \quad .$$

The *Wiener algebra* $A(J^2)$ consists of those functions $f \in \mathcal{C}_{2\pi}(J^2)$ satisfying

$$\sum_{k=-\infty}^{\infty} \sum_{l=-\infty}^{\infty} |(f,e_{k,l})| < \infty \quad .$$

It is easily seen that

$$A(J)\otimes A(J) \subseteq A(J^2) \quad .$$

The product operator $R'R''$ is given by

$$R'R''(f) = ((b_1+e_0)\otimes(b_1+e_0))*f$$

where $f \in \mathcal{C}(J^2)$. This implies the relation

$$(3) \qquad R'R''(e_{k,l}) = \lambda_k\lambda_l e_{k,l} \qquad (k,l \in \mathbb{Z})$$

with

$$\lambda_k = \frac{1}{i\cdot k} \qquad (k \in \mathbb{Z} , k \neq 0) , \quad \lambda_0 = 1.$$

Thus, $R'R''$ is a bounded injective linear operator on $\mathcal{C}(J^2)$ with the property that

$$\mathcal{R}(R'R'') \subseteq A(J^2) \quad .$$

As in the univariate case the *bivariate Korobov space* $E^a(J^2)$ is defined by the asymptotic behaviour of the bivariate Fourier coefficients :

$$E^a(J^2) = \{ f \in L^2(J^2) : (f,e_{k,l}) = \mathcal{O}(|\lambda_k\lambda_l|^a) \ (|k|,|l| \to \infty) \} .$$

Proposition 1. *If* $f,g \in E^a(J)$, *then*

(4) $$f \otimes g \in E^a(J^2) \quad .$$

Proof: We have

$$(f,e_k) = \mathcal{O}(|\lambda_k|^a) \ (|k| \to \infty) \quad , \quad (g,e_l) = \mathcal{O}(|\lambda_l|^a) \ (|l| \to \infty) \quad .$$

An application of relation (1) yields

$$(f \otimes g,e_{k,l}) = \mathcal{O}(|\lambda_k \lambda_l|^a) \ (|k|,|l| \to \infty) \quad .$$

This completes the proof of Proposition 1.

Proposition 2. *If* $f \in E^a(J^2)$ *and* $g \in \mathcal{C}(J^2)$, *then*

(5) $$f*g \in E^a(J^2) \cap A(J^2) \quad .$$

Proof: It follows from $g \in \mathcal{C}(J^2)$ that

$$\sum_{k=-\infty}^{\infty} \sum_{l=-\infty}^{\infty} |(g,e_{k,l})|^2 < \infty \quad .$$

An application of relation (2) yields

$$f*g \in E^a(J^2) \quad .$$

Moreover, we have

$$\sum_{k=-\infty}^{\infty} \sum_{l=-\infty}^{\infty} |(f*g,e_{k,l})| = \sum_{k=-\infty}^{\infty} \sum_{l=-\infty}^{\infty} |(f,e_{k,l})| \cdot |(g,e_{k,l})|$$

$$\leq \left(\sum_{k=-\infty}^{\infty} \sum_{l=-\infty}^{\infty} |(f,e_{k,l})|^2 \right)^{1/2} \cdot \left(\sum_{k=-\infty}^{\infty} \sum_{l=-\infty}^{\infty} |(g,e_{k,l})|^2 \right)^{1/2} < \infty$$

which implies

$$f*g \in A(J^2) \quad .$$

This completes the proof of Proposition 2 in view of relation (2).

Proposition 3. *If* $f \in E^a(J^2)$ *and* $g \in E^b(J^2)$, *then*

(6) $$f*g \in E^{a+b}(J^2) \quad .$$

Proof: We have

$$(f,e_{k,l}) = \mathcal{O}(|\lambda_k\lambda_l|^a) \quad (|k|,|l| \to \infty), \quad (g,e_{k,l}) = \mathcal{O}(|\lambda_k\lambda_l|^b) \quad (|k|,|l| \to \infty).$$

Taking into account relation (2) we obtain

$$(f*g,e_{k,l}) = \mathcal{O}(|\lambda_k\lambda_l|^{a+b}) \ (\ |k|,|l| \to \infty \ )$$

which implies $f*g \in E^{a+b}(J^2)$. This completes the proof of Proposition 3.

The parametric extensions P' and P'' of the projector P are given by

$$P'(f)(x,y) = \frac{f(0,y) - f(2\pi,y)}{2\pi} b_1(x) \ , \quad P''(f)(x,y) = \frac{f(x,0) - f(x,2\pi)}{2\pi} b_1(y).$$

P' and P'' are bounded linear projectors on $\mathcal{C}(J^2)$ which commute. The product and Boolean sum are given by

$$P'P''(f)(x,y) = \frac{f(0,0) - f(0,2\pi) - f(2\pi,0) - f(2\pi,2\pi)}{(2\pi)^2} b_1(x)b_1(y),$$

$$P'\oplus P''(f)(x,y) = \frac{f(0,y) - f(2\pi,y)}{2\pi} b_1(x) + \frac{f(x,0) - f(x,2\pi)}{2\pi} b_1(y)$$

$$- \frac{f(0,0) - f(0,2\pi) - f(2\pi,0) - f(2\pi,2\pi)}{(2\pi)^2} b_1(x)b_1(y).$$

It was shown in Section 4.4 that the operator $R'R''$ and the projector $P'\oplus P''$ define a right invertible operator T such that $R'R''$ is the right inverse of T with initial projector $P'\oplus P''$.

We have

$$(7) \qquad \mathscr{C}^{1,1}(J^2) = \{ f \in \mathcal{D}(T) : P'(f) \in \mathcal{D}(A'') , P''(f) \in \mathcal{D}(A') \} .$$

The right invertible operators A' and A'' are given by

$$A'(f)(x,y) = D^{1,0}f(x,y) + \frac{1}{2\pi}\int_0^{2\pi} (f(s,y) - D^{1,0}f(s,y))\, ds ,$$

$$A''(f)(x,y) = D^{0,1}f(x,y) + \frac{1}{2\pi}\int_0^{2\pi} (f(x,t) - D^{0,1}f(x,t))\, dt .$$

The operators A' and A'' commute on $\mathscr{C}^{1,1}(J^2)$. Moreover we have

$$f - P'(f) = R'A'(f) , \quad f - P''(f) = R''A''(f) .$$

The operators $R'A'$ and $R''A''$ are commuting projectors on $\mathscr{C}^{1,1}(J^2)$ which are given by

$$R'A'(f)(x,y) = \frac{1}{2\pi}\int_0^{2\pi} b_1(x-s)\cdot D^{1,0}f(s,y)\, ds + \frac{1}{2\pi}\int_0^{2\pi} f(s,y)\, ds ,$$

$$R''A''(f)(x,y) = \frac{1}{2\pi}\int_0^{2\pi} b_1(y-t)\cdot D^{0,1}f(x,t)\, dt + \frac{1}{2\pi}\int_0^{2\pi} f(x,t)\, dt .$$

In particular we have

$$(8) \qquad R'A'R''A''(f) = R''A''R'A'(f) \quad (f \in \mathscr{C}^{1,1}(J^2)) ,$$

$$(9) \qquad f - P'\oplus P''(f) = R'A'R''A''(f) \quad (f \in \mathscr{C}^{1,1}(J^2)) .$$

Proposition 4. *If* $f \in \mathscr{C}^{1,1}(J^2)$, *then*

(10) $$f - P'\oplus P''(f)$$

$$= (b_1 \otimes b_1)*D^{1,1}f + (b_1 \otimes e_0)*D^{1,0}f + (e_0 \otimes b_1)*D^{0,1}f + (e_0 \otimes e_0)*f \ .$$

Proof: Recall that

$$R''A''(f)(x,y) = \frac{1}{2\pi}\int_0^{2\pi} b_1(y-t)\cdot D^{0,1}f(x,t)\ dt + \frac{1}{2\pi}\int_0^{2\pi} f(x,t)\ dt \ ,$$

$$R'A'(f)(x,y) = \frac{1}{2\pi}\int_0^{2\pi} b_1(x-s)\cdot D^{1,0}f(s,y)\ ds + \frac{1}{2\pi}\int_0^{2\pi} f(s,y)\ ds \ .$$

Then we obtain

$$f - P'\oplus P''(f) = R'A'R''A''(f)$$

$$= \frac{1}{(2\pi)^2}\int_0^{2\pi}\int_0^{2\pi} b_1(x-s)b_1(y-t)D^{1,1}f(s,t)\ ds\ dt + \frac{1}{(2\pi)^2}\int_0^{2\pi}\int_0^{2\pi} f(s,t)dsdt$$

$$+ \frac{1}{(2\pi)^2}\int_0^{2\pi}\int_0^{2\pi} b_1(y-t)D^{0,1}f(s,t)dsdt + \frac{1}{(2\pi)^2}\int_0^{2\pi}\int_0^{2\pi} b_1(x-s)D^{1,0}f(s,t)dsdt$$

which completes the proof of Proposition 4 .

Proposition 5. *The inclusion*

(11) $$\mathscr{C}^{1,1}(J^2) \subseteq E^1(J^2)$$

is valid.

Proof: We have

$$f(\cdot,0)\ ,\ f(\cdot,2\pi)\ ,\ f(0,\cdot)\ ,\ f(2\pi,\cdot) \quad \in \quad \mathcal{C}^1(J) \quad .$$

It follows from Proposition 2 of Section 7.1 and Proposition 1 that

$$P'\oplus P''(f) \quad \in \quad E^1(J^2) \quad .$$

Taking into account relations (4), (5), and (10) we obtain

$$R'A'R''A''(f) \quad \in \quad E^1(J^2) \quad .$$

This completes the proof of Proposition 5.

Proposition 6. *The inclusion*

(12) $$\mathcal{C}^{2,2}(J^2) \cap \mathcal{C}_{2\pi}(J^2) \quad \subseteq \quad E^2(J^2)$$

is valid.

Proof: Note first that

(13) $$\mathcal{N}(P'\oplus P'') \quad = \quad \mathcal{C}_{2\pi}(J^2) \quad .$$

Taking into account relation (10) we obtain

$$f \quad = \quad (b_1\otimes b_1)*D^{1,1}f \ +\ (b_1\otimes e_0)*D^{1,0}f \ +\ (e_0\otimes b_1)*D^{0,1}f \ +\ (e_0\otimes e_0)*f$$

for f $\mathcal{C}^{2,2}(J^2) \cap \mathcal{C}_{2\pi}(J^2)$. It follows from Proposition 5 that

$$D^{1,1}f\ ,\ D^{1,0}f\ ,\ D^{0,1}f \quad \in \quad E^1(J^2) \quad .$$

It follows from Proposition 2 that

$$b_1\otimes b_1\ ,\ b_1\otimes e_0\ ,\ e_0\otimes b_1 \quad \in \quad E^1(J^2) \quad .$$

An application of Proposition 3 completes the proof of Proposition 6 .

Proposition 7. *If* $f \in \mathscr{C}^{m,m}(J^2)$, *then the equality*

(14) $$f = P'_m \oplus P''_m(f) + R'^m A'^m R''^m A''^m(f)$$

holds with

(15) $$P'_m \oplus P''_m(f)(x,y)$$

$$= \sum_{j=0}^{m-1} \frac{D^{j,0}f(0,y) - D^{j,0}f(2\pi,y)}{2\pi} b_{j+1}(x)$$

$$+ \sum_{k=0}^{m-1} \frac{D^{0,k}f(x,0) - D^{0,k}f(x,2\pi)}{2\pi} b_{k+1}(y)$$

$$- \sum_{j=0}^{m-1} \sum_{k=0}^{m-1} \frac{1}{(2\pi)^2} \Big(D^{j,k}f(0,0) - D^{j,k}f(0,2\pi) - D^{j,k}f(2\pi,0)$$

$$+ D^{j,k}f(2\pi,2\pi)\Big) \cdot b_{j+1}(x) \cdot b_{k+1}(y)$$

and

(16) $$R'^m A'^m R''^m A''^m(f)$$

$$= (b_m \otimes b_m)*D^{m,m}f + (b_m \otimes e_0)*D^{m,0}f + (e_0 \otimes b_m)*D^{0,m}f + (e_0 \otimes e_0)*f .$$

Proof: Recall that

$$P_m(f) = \sum_{j=0}^{m-1} \frac{D^j f(0) - D^j f(2\pi)}{2\pi} b_{j+1}(x), \qquad R^m A^m(f) = b_m * D^m f + e_0 * f.$$

Relation (14) follows from Propositions 1 and 2 of Section 4.5 .The representation formulas (15) and (16) are obtained by the method of parametric extension as in the proof of Proposition 4 .

Proposition 8. *The inclusion*

$$\mathscr{C}^{m+1,m+1}(J^2) \cap \mathscr{C}_{2\pi}^{m-1,m-1}(J^2) \subseteq E^{m+1}(J^2) \tag{17}$$

is valid for every $m \in \mathbb{N}$.

Proof: It follows from relation (15) that

$$\mathscr{N}(P'_m \oplus P''_m) = \mathscr{C}_{2\pi}^{m-1,m-1}(J^2) \quad . \tag{18}$$

Taking into account relation (14) we obtain

$$f = (b_m \otimes b_m)^* D^{m,m} f + (b_m \otimes e_0)^* D^{m,0} f + (e_0 \otimes b_m)^* D^{0,m} f + (e_0 \otimes e_0)^* f \ .$$

By assumption we have

$$D^{m,m} f \ , \ D^{m,0} f \ , \ D^{0,m} f \quad \in \quad \mathscr{C}^{1,1}(J^2) \quad .$$

Using Propositions 5 and 2 we obtain $f \in E^{m+1}(J^2)$. This completes the proof of Proposition (8) .

7.4 Fourier tensor product approximation

The projector of Fourier tensor product approximation is obtained as the product of parametrically extended Fourier partial sum projectors. Its approximation power is determined in bivariate Korobov spaces. Using Boolean sum techniques a bivariate analogue of the Krylov-Lanczos approximant is constructed.

The univariate Fourier projector on $E^a(J)$ is given by

$$F_m(f) = \sum_{k=-m}^{m} (f,e_k)\cdot e_k .$$

The parametric extensions F'_m and F''_n are projectors on $E^a(J^2)$ which possess the representatons

$$F'_m(f) = \sum_{j=-m}^{m} \sum_{k=-\infty}^{\infty} (f,e_{j,k})e_{j,k}, \quad F''_n(f) = \sum_{j=-\infty}^{\infty} \sum_{k=-n}^{n} (f,e_{j,k})e_{j,k}.$$

It is easily seen that F'_m and F''_n are commuting projectors on $E^a(J^2)$. The *Fourier tensor product projector* is given by

$$F'_m F''_n(f) = \sum_{j=-m}^{m} \sum_{k=-n}^{n} (f,e_{j,k})\cdot e_{j,k} .$$

Proposition 1. *If* $f \in E^a(J^2)$, *then*

(1) $$\|f - F'_m F''_n(f)\| = \mathcal{O}(m^{-a+1/2} + n^{-a+1/2}) \quad (m \to \infty,\ n \to \infty) .$$

Proof: The remainder projectors F'^c_m and F''^c_n satisfy

$$F'^c_m(f) = \sum_{|j|>m} \sum_{k=-\infty}^{\infty} (f,e_{j,k})e_{j,k}, \quad F''^c_n(f) = \sum_{j=-\infty}^{\infty} \sum_{|k|>n} (f,e_{j,k})e_{j,k}.$$

This implies

(2) $$\|F'^{c}_{m}(f)\|^2 = \mathcal{O}(m^{-2a+1}) \quad (m \to \infty) \ ,$$

(3) $$\|F'^{c}_{n}(f)\|^2 = \mathcal{O}(n^{-2a+1}) \quad (n \to \infty) \ .$$

The product of the remainder projectors is

$$F'^{c}_{m}F''^{c}_{n}(f) = \sum_{|j|>m} \sum_{|k|>n} (f,e_{j,k})e_{j,k}.$$

Thus we have

(4) $$\|F'^{c}_{m}F''^{c}_{n}(f)\|^2 = \mathcal{O}(m^{-2a+1}\cdot n^{-2a+1}) \quad (m \to \infty \ , \ n \to \infty) \ .$$

The parametrically extended Fourier projectors

$$F'_{m} \ , \ F''_{n} \quad (m,n \in \mathbb{Z}_{+})$$

generate a Boolean algebra $\mathfrak{F}$ of projectors on $E^{a}(J^2)$. The duality relations in Boolean algebras yield

(5) $$(F'_{m}F''_{n})^{c} = F'^{c}_{m} \oplus F''^{c}_{n} \ .$$

Then we obtain $(F'_{m}F''_{n})^{c} = F'^{c}_{m} + F''^{c}_{n} - F'^{c}_{m}F''^{c}_{n}$ which implies

(6) $$(F'_{m}F''_{n})^{c} = F'^{c}_{m}F''_{n} + F'_{m}F''^{c}_{n} + F'^{c}_{m}F''^{c}_{n}.$$

Relation (6) is the *orthogonal decomposition* of the remainder of the Fourier tensor product projector $F'_{m}F''_{n}$. Then we can conclude

$$\|f - F'_{m}F''_{n}(f)\|^2 = \|F'^{c}_{m}F''_{n}(f)\|^2 + \|F'_{m}F''^{c}_{n}(f)\|^2 + \|F'^{c}_{m}F''^{c}_{n}(f)\|^2$$

$$\leq \|F'^{c}_{m}(f)\|^2 + \|F''^{c}_{n}(f)\|^2 + \|F'^{c}_{m}F''^{c}_{n}(f)\|^2 \ .$$

Taking into account the relations (2), (3), and (4) we obtain

(7) $$\|f - F'_m F''_n(f)\|^2 = \mathcal{O}(m^{-2a+1} + n^{-2a+1}) \quad (m \to \infty, n \to \infty) .$$

This completes the proof of Proposition 1.

Proposition 2. *If* $f \in \mathcal{C}^{1,1}(J^2)$, *then*

(8) $$\|f - F'_m F''_n(f)\| = \mathcal{O}(m^{-1/2} + n^{-1/2}) \quad (m \to \infty, n \to \infty) .$$

Proof: It follows from Proposition 5 of Section 7.3 that $f \in E^1(J^2)$. An application of Proposition 1 yields relation (8) . This completes the proof of Proposition 2.

Proposition 3. *If* $f \in \mathcal{C}^{q+1,q+1}(J^2) \cap \mathcal{C}^{q-1,q-1}_{2\pi}(J^2)$, *then*

(9) $$\|f - F'_m F''_n(f)\| = \mathcal{O}(m^{-q-1/2} + n^{-q-1/2}) \quad (m \to \infty, n \to \infty) .$$

Proof: It follows from Proposition 5 of Section 7.3 that $f \in E^{q+1}(J^2)$. An application of Proposition 1 yields relation (9). This completes the proof of Proposition 3.

We can derive error estimates for Fourier tensor product projector $F'_m F''_n$ in the uniform norm for the case $a > 1$.

Proposition 4. *If* $f \in E^a(J^2)$ *with* $a > 1$, *then*

(10) $$\|f - F'_m F''_n(f)\|_\infty = \mathcal{O}(m^{-a+1} + n^{-a+1}) \quad (m \to \infty, n \to \infty) .$$

Proof: Note first that

$$E^a(J^2) \subseteq A(J^2) \qquad (a > 1) .$$

Then we have

$$\|F'^c_m F''^c_n(f)\|_\infty \le \sum_{|j|>m} \sum_{|k|>n} |(f,e_{j,k})| ,$$

$$\|F'^c_m F''_n(f)\|_\infty \le \sum_{|j|>m} \sum_{|k|\le n} |(f,e_{j,k})|, \qquad \|F'_m F''^c_n(f)\|_\infty \le \sum_{|j|\le m} \sum_{|k|>n} |(f,e_{j,k})|.$$

Thus we obtain

$$\|F'^c_m F''^c_n(f)\|_\infty = \mathcal{O}(m^{-a+1}\cdot n^{-a+1}) \quad (m \to \infty , n \to \infty) ,$$

$$\|F'^c_m F''_n(f)\|_\infty = \mathcal{O}(m^{-a+1}) , \qquad \|F'_m F''^c_n(f)\|_\infty = \mathcal{O}(n^{-a+1}) \quad (m \to \infty , n \to \infty).$$

Then we can conclude

$$\begin{aligned} \|f - F'_m F''_n(f)\|_\infty &\le \|F'^c_m F''_n(f)\|_\infty + \|F'_m F''^c_n(f)\|_\infty + \|F'^c_m F''^c_n(f)\|_\infty \\ &= \mathcal{O}(m^{-a+1} + n^{-a+1}) \quad (m \to \infty, n \to \infty) . \end{aligned}$$

This completes the proof of Proposition 4.

Proposition 5. *If* $f \in \mathscr{C}^{q+1,q+1}(J^2) \cap \mathscr{C}^{q-1,q-1}_{2\pi}(J^2)$, *then*

(11) $$\|f - F'_m F''_n(f)\|_\infty = \mathcal{O}(m^{-q} + n^{-q}) \quad (m \to \infty, n \to \infty) .$$

Proof: It follows from Proposition 5 of Section 7.3 that $f \in E^{q+1}(J^2)$. An application of Proposition 4 yields relation (11). This completes the proof of Proposition 5.

For $f \in \mathscr{C}^{q+1,q+1}(J^2)$ we introduce the *bivariate Krylov-Lanczos approximant* of f which is defined by

$$K^q_{m,n}(f) := P'_q \oplus P''_q(f) + F'_m F''_n(f - P'_q \oplus P''_q(f))$$

with

$$P'_q \oplus P''_q(f)(x,y)$$

$$= \sum_{j=0}^{q-1} \frac{D^{j,0}f(0,y)-D^{j,0}f(2\pi,y)}{2\pi} b_{j+1}(x) + \sum_{k=0}^{q-1} \frac{D^{0,k}f(x,0)-D^{0,k}f(x,2\pi)}{2\pi} b_{k+1}(y)$$

$$- \sum_{j=0}^{q-1} \sum_{k=0}^{q-1} \frac{D^{j,k}f(0,0)-D^{j,k}f(0,2\pi)-D^{j,k}f(2\pi,0)+D^{j,k}f(2\pi,2\pi)}{(2\pi)^2} b_{j+1}(x) b_{k+1}(y).$$

Proposition 6. *If* $f \in \mathscr{C}^{q+1,q+1}(J^2)$ *, then*

(12) $$\|f - K^q_{m,n}(f)\| = \mathcal{O}(m^{-q-1/2} + n^{-q-1/2}) \quad (m \to \infty,\ n \to \infty) .$$

Proof: It follows Proposition 8 of Section 7.3 that

(13) $$f - P'_q \oplus P''_q(f) \in E^{q+1}(J^2) .$$

Since

(14) $$f - K^q_{m,n}(f) = f - P'_q \oplus P''_q(f) - F'_m F''_n(f - P'_q \oplus P''_q(f)) ,$$

relation (12) follows from Proposition 3 . This completes the proof of Proposition 6.

Proposition 7. *If* $f \in \mathscr{C}^{q+1,q+1}(J^2)$, *then*

(15) $$\|f - K^q_{m,n}(f)\|_\infty = \mathcal{O}(m^{-q} + n^{-q}) \quad (m \to \infty,\ n \to \infty) .$$

Proof: Relation (15) follows from (13) , (14) , and Proposition 5. This completes the proof of Proposition 7.

7.5 Fourier blending approximation

The transfinite analogue of the Fourier tensor product projector $F'_m F''_n$ is the *Fourier blending projector* $F'_m \oplus F''_n$:

$$F'_m \oplus F''_n(f)$$

$$= \sum_{j=-m}^{m} \sum_{k=-\infty}^{\infty} (f,e_{j,k})e_{j,k} + \sum_{j=-\infty}^{\infty} \sum_{k=-n}^{n} (f,e_{j,k})e_{j,k} - \sum_{j=-m}^{m} \sum_{k=-n}^{n} (f,e_{j,k})e_{j,k} .$$

$F'_m \oplus F''_n$ is an element of the Boolean algebra $\mathfrak{F}$ of bivariate Fourier projectors.

Proposition 1. *If* $f \in E^a(J^2)$, *then*

(1) $$\|f - F'_m \oplus F''_n(f)\| = \mathcal{O}(m^{-a+1/2} \cdot n^{-a+1/2}) \quad (m \to \infty,\ n \to \infty) .$$

Proof: It follows from the duality relations of the Boolean algebra $\mathfrak{F}$ that the relation

(2) $$(F'_m \oplus F''_n)^c = F'^c_m F''^c_n$$

holds . Relation (1) follows now from relation (4) of Section 7.4 .

Proposition 2. *If* $f \in \mathcal{C}^{1,1}(J^2)$, *then*

(3) $$\|f - F'_m \oplus F''_n(f)\| = \mathcal{O}(m^{-1/2} \cdot n^{-1/2}) \quad (m \to \infty,\ n \to \infty) .$$

Proof: It follows from Proposition 5 of Section 7.3 that $f \in E^1(J^2)$. An application of Proposition 1 yields relation (3) . This completes the proof of Proposition 2.

Proposition 3. *If* $f \in \mathcal{C}^{q+1,q+1}(J^2) \cap \mathcal{C}_{2\pi}^{q-1,q-1}(J^2)$, *then*

$$(4) \qquad \|f - F'_m \oplus F''_n(f)\| = \mathcal{O}(m^{-q-1/2} \cdot n^{-q-1/2}) \quad (m \to \infty,\ n \to \infty) .$$

Proof: It follows from Proposition 5 of Section 7.3 that $f \in E^{q+1}(J^2)$. An application of Proposition 1 yields relation (4). This completes the proof of Proposition 3.

We can derive error estimates for the Fourier blending projector $F'_m \oplus F''_n$ in the uniform norm for $a > 1$.

Proposition 4. *If* $f \in E^a(J^2)$ *with* $a > 1$, *then*

$$(5) \qquad \|f - F'_m \oplus F''_n(f)\|_\infty = \mathcal{O}(m^{-a+1} \cdot n^{-a+1}) \quad (m \to \infty,\ n \to \infty) .$$

Proof: Recall that

$$E^a(J^2) \subseteq A(J^2) \qquad (\ a > 1\ ) .$$

Taking into account relation (2) we obtain

$$\|f - F'_m \oplus F''_n(f)\|_\infty \le \sum_{|j|>m} \sum_{|k|>n} |(f,e_{j,k})| = \mathcal{O}(m^{-a+1} \cdot n^{-a+1}) \quad (m \to \infty,\ n \to \infty)$$

which completes the proof of Proposition 4.

Proposition 5. *If* $f \in \mathcal{C}^{q+1,q+1}(J^2) \cap \mathcal{C}_{2\pi}^{q-1,q-1}(J^2)$, *then*

$$(6) \qquad \|f - F'_m \oplus F''_n(f)\|_\infty = \mathcal{O}(m^{-q} \cdot n^{-q}) \quad (m \to \infty,\ n \to \infty) .$$

Proof: It follows from Proposition 5 of Section 7.3 that $f \in E^{q+1}(J^2)$. An application of Proposition 4 yields relation (6). This completes the proof of Proposition 5.

Remark 1. The Krylov-Lanczos approximant can be defined for the Fourier blending projector in the same way as for the Fourier tensor product projector . For this approximant error estimates can be derived which are similar to (4) and (6) . We omit the details.

7.6 Discrete Fourier blending approximation

In Section 7.4 and in Section 7.5 we have determined the approximation power of the Fourier product projector $F'_nF''_n$ and the Fourier blending projector $F'_m\oplus F''_m$. We have shown that

$$\|f - F'_nF''_n(f)\| = \mathcal{O}(n^{-a+1/2}) \quad (n \to \infty), \tag{1}$$

$$\|f - F'_m\oplus F''_m(f)\| = \mathcal{O}(m^{-2a+1}) \quad (m \to \infty) \tag{2}$$

where $f \in E^a(J^2)$. Choosing $n = m^2$ in relation (1) it follows that the Fourier tensor product projector $F'_{m^2}\cdot F''_{m^2}$ and the Fourier blending projector $F'_m\oplus F''_m$ possess the same approximation power in $E^a(J^2)$. To reduce the dimension of $\mathcal{R}(F'_{m^2}\cdot F''_{m^2})$ we introduce the *projector of discrete Fourier blending approximation* which is defined by

$$\mathbf{F}^m_2 := F'_m\cdot F''_{m^2} \oplus F'_{m^2}\cdot F''_m = F'_m\cdot F''_{m^2} + F'_{m^2}\cdot F''_m - F'_m\cdot F''_m .$$

Proposition 1. *If* $f \in E^a(J^2)$, *then*

$$\|f - \mathbf{F}^m_2(f)\| = \mathcal{O}(m^{-2a+1}) \quad (m \to \infty) . \tag{3}$$

Proof: We apply Proposition 2 of Section 1.4 and obtain

$$\mathbf{F}^m_2 = (F'_m\cdot F''_{m^2} \oplus F'_{m^2}\cdot F''_m)^c = F'^c_{m^2} + F'^c_m\cdot F''^c_m + F''^c_{m^2} - (F'^c_{m^2}\cdot F''^c_m + F'^c_m\cdot F''^c_{m^2}),$$

i. e., we have

$$(F'_m\cdot F''_{m^2} \oplus F'_{m^2}\cdot F''_m)^c = F'^c_{m^2}\cdot F''_m + F'^c_m\cdot F''^c_m + F'_m\cdot F''^c_{m^2} . \tag{4}$$

Relation (4) is the *orthogonal decomposition* of the remainder projector of discrete Fourier blending approximation. Then we can conclude

$$\|(F'_m \cdot F''_m 2 \oplus F'_m 2 \cdot F''_m)^c(f)\|^2 = \|F'_m 2^c \cdot F''_m(f)\|^2 + \|F'^{\,c}_m \cdot F''^{\,c}_m(f)\|^2 + \|F'_m \cdot F''_m 2^c(f)\|^2$$

$$\leq \|F'_m 2^c(f)\|^2 + \|F'^{\,c}_m \cdot F''^{\,c}_m(f)\|^2 + \|F''_m 2^c(f)\|^2 .$$

Taking into account the relations (2), (3), and (4) of Section 7.4 we obtain the asymptotic relation (3) . This completes the proof of Proposition 1 .

Remark 1. The bivariate Fourier projectors $F'_m \cdot F''_m 2 \oplus F'_m 2 \cdot F''_m$ and $F'_m 2 \cdot F''_m 2$ possess the same approximation power in view of Proposition 1. Moreover we have

$$\dim \mathcal{R}(\mathbf{F}^m_2) = \dim R(F'_m \cdot F''_m 2 \oplus F'_m 2 \cdot F''_m) = (2m+1)(4m^2-2m+1) ,$$

$$\dim \mathcal{R}(F'_m 2 \cdot F''_m 2) = (2m^2+1)^2 .$$

Thus $F'_m \cdot F''_m 2 \oplus F'_m 2 \cdot F''_m$ is more efficient than $F'_m 2 \cdot F''_m 2$ as an approximation operator .

Proposition 2. *If* $f \in \mathcal{C}^{1,1}(J^2)$, *then*

(5) $$\|f - \mathbf{F}^m_2(f)\| = \mathcal{O}(m^{-1}) \quad (m \to \infty) .$$

Proof: It follows from Proposition 5 of Section 7.3 that $f \in E^1(J^2)$. An application of Proposition 1 yields relation (5) . This completes the proof of Proposition 2.

Proposition 3. *If* $f \in \mathcal{C}^{q+1,q+1}(J^2) \cap \mathcal{C}^{q-1,q-1}_{2\pi}(J^2)$, *then*

(6) $$\|f - \mathbf{F}^m_2(f)\| = \mathcal{O}(m^{-2q-1}) \quad (m \to \infty) .$$

Proof: It follows from Proposition 5 of Section 7.3 that $f \in E^{q+1}(J^2)$. An application of Proposition 1 yields relation (3). This completes the proof of Proposition 3.

Proposition 4. *If* $f \in E^a(J^2)$ *with* $a > 1$, *then*

$$\|f - \mathbf{F}_2^m(f)\|_\infty = \mathcal{O}(m^{-2a+2}) \quad (m \to \infty) . \tag{7}$$

Proof: Recall that

$$E^a(J^2) \subseteq A(J^2) \qquad (a > 1) .$$

Taking into account relation (4) we obtain

$$\|f - (F'_m \cdot F''_{m^2} \oplus F'_{m^2} \cdot F''_m)(f)\|_\infty$$

$$\le \|F'^{c}_{m^2} \cdot F''_m(f)\|_\infty + \|F'^{c}_m \cdot F''^{c}_m(f)\|_\infty + \|F'_m \cdot F''^{c}_{m^2}(f)\|_\infty$$

$$\le \sum_{|j|>m^2} \sum_{|k|\le m} |(f,e_{j,k})| + \sum_{|j|>m} \sum_{|k|>m} |(f,e_{j,k})| + \sum_{|j|\le m} \sum_{|k|\ge m^2} |(f,e_{j,k})|$$

$$= \mathcal{O}(m^{-2a+2}) \qquad (m \to \infty)$$

which completes the proof of Proposition 4.

Proposition 5. *If* $f \in \mathcal{C}^{q+1,q+1}(J^2) \cap \mathcal{C}^{q-1,q-1}_{2\pi}(J^2)$, *then*

$$\|f - \mathbf{F}_2^m(f)\|_\infty = \mathcal{O}(m^{-2q}) \quad (m \to \infty) . \tag{8}$$

Proof: It follows from Proposition 5 of Section 7.3 that $f \in E^{q+1}(J^2)$.

An application of Proposition 4 yields relation (8). This completes the proof of Proposition 5.

Remark 1. The Krylov–Lanczos approximant can be defined for the discrete Fourier blending projector in the same way as for the Fourier tensor product projector . For this approximant error estimates can be derived which are similar to (6) and (8) . We omit the details.

7.7 Hyperbolic Fourier approximation

The injective operator of the bivariate periodic Taylor formula possesses a point spectrum which can be described most conveniently as a biinfinite double sequence. Each absolute value of a fixed eigenvalue determines uniquely a set of eigenfunctions whose eigenvalues are greater or equal in absolute value than the fixed one. The corresponding orthogonal projector is the spectral partial sum projector or hyperbolic Fourier projector whose approximation properties will be investigated.

The operator of hyperbolic Fourier approximation is defined by

$$H_n(f) \quad = \quad \sum_{\substack{|jk| \le n \\ |j|,|k| \le n}} (f, e_{j,k}) e_{j,k} \quad .$$

It is obvious from the definition that H_n is a projector on $E^a(J^2)$. The set of *frequencies* of H_n is

$$\mathfrak{h}_n \quad := \quad \{ \ (j,k) \in \mathbb{Z}^2 \ : \ |j| \ , \ |k| \le n \ , \ |jk| \le n \ \} \quad .$$

It is a *cross hyperbolic* set of lattice points. Its number is the dimension of $\mathcal{R}(H_n)$. It is known from number theory that

$$\dim \mathcal{R}(H_n) \quad = \quad |\mathfrak{h}_n| \quad = \quad 4 \cdot n \cdot \log(n) + \mathcal{O}(n) \qquad (\ n \to \infty \ ) \quad .$$

Recall that

$$\lambda_k \quad = \quad \frac{1}{i \cdot k} \qquad (\ k \in \mathbb{Z} \ , \ k \ne 0 \ ) \quad , \quad \lambda_0 \quad = \quad 1 \quad .$$

The operator R'R" satisfies

$$R'R''(e_{j,k}) \quad = \quad \lambda_j \lambda_k \cdot e_{j,k} \qquad (\ j,k \in \mathbb{Z} \ ) \quad .$$

The hyperbolic Fourier projector H_n is the *spectral partial sum operator* of R'R" :

(2) $$H_n(f) = \sum_{|\lambda_j \lambda_k| \ge \frac{1}{n}} (f, e_{j,k}) e_{j,k} .$$

With respect to this notation the Fourier product projector $F'_n F''_n$ is given by

(3) $$F'_n F''_n(f) = \sum_{|\lambda_j| \ge \frac{1}{n}} \sum_{|\lambda_k| \ge \frac{1}{n}} (f, e_{j,k}) e_{j,k} .$$

We will show that the hyperbolic Fourier projector H_n possesses the same approximation power up to a logarithmic factor as the Fourier product projector $F'_n F''_n$. Since

$$\dim \mathcal{R}(F'_n F''_n) = 4 \cdot n^2 + 4 \cdot n + 1 ,$$

$$\dim \mathcal{R}(H_n) = 4 \cdot n \cdot \log(n) + \mathcal{O}(n) \quad (n \to \infty) ,$$

it follows that H_n is much more efficient then $F'_n F''_n$ as approximation operator.
The remainder projector of H_n satisfies

(4) $$H^c_n(f) = \sum_{|jk| \ge n} (f, e_{j,k}) e_{j,k} + \sum_{|j| \ge n} (f, e_{j,0}) e_{j,0} + \sum_{|k| \ge n} (f, e_{0,k}) e_{0,k} .$$

It is well known that

(5) $$\sum_{j=n+1}^{\infty} j^{-b} = \mathcal{O}(n^{-b+1}) \quad (n \quad \infty)$$

where $b > 1$. We will use a similar result for the double series

$$\sum_{\substack{jk > n \\ j > 1 , k > 1}} (jk)^{-b} .$$

Lemma 1. *If* $j\cdot k > n,\ j > 1,\ k > 1$ *with* $n \geq 4$, *then*

(6) $$(j-1)(k-1) \geq \frac{n}{2} - 1 .$$

Proof: We have $j\cdot k = N$ with $N > n$. Then we obtain

$$(j-1)(k-1) = N + 1 - (k + \frac{N}{k})$$

with $2 \leq k \leq N/2$ for simplicity. We introduce the function

$$\phi(x) = N + 1 - (x + \frac{N}{x}) \quad (2 \leq x \leq N/2) .$$

It is easily seen that

$$\phi(x) \geq \phi(2) \quad (2 \leq x \leq N/2) .$$

Now we can conclude

$$(j-1)(k-1) \geq \frac{N}{2} - 1 > \frac{n}{2} - 1 .$$

This completes the proof of Lemma 1.

Lemma 2. *If* $n \geq 4$ *and* $b > 1$, *then*

(7) $$\sum_{\substack{jk>n \\ j>1,\, k>1}} (jk)^{-b} \leq \iint\limits_{\substack{xy\geq -1+n/2 \\ x,y\geq 1}} (xy)^{-b}\, dx\, dy .$$

Proof: Taking into account Lemma 1 we obtain

$$\sum_{\substack{jk>n \\ j>1,\, k>1}} (jk)^{-b} = \sum_{\substack{jk>n \\ j>1,\, k>1}} (jk)^{-b}\cdot(j-(j-1))\cdot(k-(k-1))$$

$$\leq \sum_{\substack{jk>n \\ j>1,\, k>1}} \int_{j-1}^{j}\int_{k-1}^{k} (xy)^{-b} dxdy \leq \iint\limits_{\substack{xy\geq -1+n/2 \\ x,y\geq 1}} (xy)^{-b}\, dxdy$$

which completes the proof of Lemma 3.

Lemma 3. *If* $b > 1$, *then*

(8) $$\iint\limits_{\substack{xy \geq -1+n/2 \\ x,y\geq 1}} (xy)^{-b}\, dx\, dy = \mathcal{O}(n^{-b+1}\cdot \log(n)) \quad (n \to \infty) .$$

Proof: Let $N = \frac{n}{2} - 1$. Then we have

$$\iint\limits_{\substack{xy\geq N \\ x,y\geq 1}} (xy)^{-b}\, dx\, dy$$

$$= \int_1^{\infty}\int_1^{\infty} (xy)^{-b}\, dx\, dy - \int_1^{N}\int_1^{N/x} (xy)^{-b}\, dx\, dy$$

$$= \left(\int_1^{\infty} x^{-b}\, dx \right)^2 - \int_1^{N} x^{-b} \left(\int_1^{N/x} y^{-b}\, dy \right) dx$$

$$= (b-1)^{-2} - \int_1^{N} x^{-b} \left((-b+1)^{-1}\left(N^{-b+1}\cdot x^{b-1} - 1\right)\right) dx$$

$$= (b-1)^{-2} + (b-1)^{-1}\cdot N^{-b+1} \int_1^{N} x^{-1}\, dx - (b-1)^{-1} \int_1^{N} x^{-b}\, dx$$

$$= (b-1)^{-2} + (b-1)^{-1}\cdot N^{-b+1}\cdot \log(N) + (b-1)^{-2}(N^{-b+1}-1) ,$$

$$= (b-1)^{-1}\cdot N^{-b+1}\cdot \log(N) + (b-1)^{-2}\cdot N^{-b+1}$$

This completes the proof of Lemma 3.

Proposition 1. *If* $f \in E^a(J^2)$, *then*

(9) $$\|f - H_n(f)\| = \mathcal{O}(n^{-a+1/2} \cdot \log(n)^{1/2}) \quad (n \to \infty).$$

Proof: Using relation (4) and Parseval's equality we get

$$\|f - H_n(f)\|^2 = \sum_{|jk| \geq n} |(f, e_{j,k})|^2 + \sum_{|j| \geq n} |(f, e_{j,0})|^2 + \sum_{|k| \geq n} |(f, e_{0,k})|^2 .$$

It follows from $f \in E^a(J^2)$ that

$$\sum_{|j| \geq n} |(f, e_{j,0})|^2 = \mathcal{O}(n^{-2a+1}), \qquad \sum_{|k| \geq n} |(f, e_{0,k})|^2 = \mathcal{O}(n^{-2a+1}).$$

Moreover we have

$$\sum_{|jk| \geq n} |(f, e_{j,k})|^2 = \mathcal{O}\Big(\sum_{\substack{jk > n \\ j>1,\, k>1}} (jk)^{-2a} \Big) + \mathcal{O}(n^{-2a+1}) .$$

In view of Lemma 2 and Lemma 3 we can conclude

$$\|f - H_n(f)\|^2 = \mathcal{O}(n^{-2a+1}) + \mathcal{O}(n^{-2a+1} \cdot \log(n)) = \mathcal{O}(n^{-2a+1} \cdot \log(n)) \quad (n \to \infty)$$

which completes the proof of Proposition 1.

Proposition 2. *If* $f \in \mathcal{C}^{1,1}(J^2)$, *then*

(10) $$\|f - H_n(f)\| = \mathcal{O}(n^{-1/2} \cdot \log(n)^{1/2}) \quad (n \to \infty).$$

Proof: It follows from Proposition 5 of Section 7.3 that $f \in E^1(J^2)$. An application of Proposition 1 yields relation (10) . This completes the proof of Proposition 2.

Proposition 3. *If* $f \in \mathscr{C}^{q+1,q+1}(J^2) \cap \mathscr{C}_{2\pi}^{q-1,q-1}(J^2)$, *then*

(11) $$\|f - H_n(f)\| = \mathcal{O}(n^{-q-1/2} \cdot \log(n)^{1/2}) \quad (n \to \infty) .$$

Proof: It follows from Proposition 5 of Section 7.3 that $f \in E^{q+1}(J^2)$. An application of Proposition 1 yields relation (11). This completes the proof of Proposition 3.

Proposition 4. *If* $f \in E^a(J^2)$ *with* $a > 1$, *then*

(12) $$\|f - H_n(f)\|_\infty = \mathcal{O}(n^{-a+1} \cdot \log(n)) \quad (n \to \infty) .$$

Proof: Recall that

$$E^a(J^2) \subseteq A(J^2) \quad (a > 1) .$$

Taking into account relation (4) we obtain

$$\|f - H_n(f)\|_\infty \le \sum_{|jk| \ge n} |(f,e_{j,k})| + \sum_{|j| \ge n} |(f,e_{j,0})| + \sum_{|k| \ge n} |(f,e_{0,k})| .$$

It follows from $f \in E^a(J^2)$ that

$$\sum_{|j| \ge n} |(f,e_{j,0})| = \mathcal{O}(n^{-a+1}), \qquad \sum_{|k| \ge n} |(f,e_{0,k})| = \mathcal{O}(n^{-a+1}).$$

Moreover we have

$$\sum_{|jk| \ge n} |(f,e_{j,k})| = \mathcal{O}\Big(\sum_{\substack{jk > n \\ j > 1,\, k > 1}} (jk)^{-a} \Big) + \mathcal{O}(n^{-a+1}) .$$

In view of Lemma 2 and Lemma 3 we can conclude

$$\|f - H_n(f)\|_\infty = \mathcal{O}(n^{-a+1}) + \mathcal{O}(n^{-a+1}\cdot\log(n)) = \mathcal{O}(n^{-a+1}\cdot\log(n)) \quad (n \to \infty)$$

which completes the proof of Proposition 1.

Proposition 5. *If* $f \in \mathscr{C}^{q+1,q+1}(J^2) \cap \mathscr{C}_{2\pi}^{q-1,q-1}(J^2)$, *then*

(13) $$\|f - H_n(f)\|_\infty = \mathcal{O}(n^{-q}\cdot\log(n)) \quad (n \to \infty) .$$

Proof: It follows from Proposition 5 of Section 7.3 that $f \in E^{q+1}(J^2)$. An application of Proposition 4 yields relation (13). This completes the proof of Proposition 5.

Remark 1. The Krylov-Lanczos approximant can be defined for the hyperbolic Fourier projector in the same way as for the Fourier product projector . For this approximant error estimates can be derived which are similar to (11) and (13) . We omit the details.

7.8 Pseudo-hyperbolic Fourier approximation

We use Boolean methods to construct a bivariate Fourier approximation projector which possesses the same approximation power as the corresponding hyperbolic Fourier projector . Let m and r be positive integers. The projector of *pseudo-hyperbolic Fourier approximation* is an element of the Boolean algebra $\mathfrak{F}$ of bivariate Fourier projectors which is given by

$$\mathbf{F}_r^m \;=\; F'_{m^0} \cdot F''_{m^r} \oplus F'_{m^1} \cdot F''_{m^{r-1}} \oplus \cdots \oplus F'_{m^{r-1}} \cdot F''_{m^1} \oplus F'_{m^r} \cdot F''_{m^0} \;.$$

It is easily seen that $\mathbf{F}_r^m$ possesses the representation

$$\text{(1)} \qquad \mathbf{F}_r^m(f) \;=\; \sum_{\substack{u+v=r \\ |j| \le m^u \\ |k| \le m^v}} (f, e_{j,k}) e_{j,k} \quad .$$

It is obvious from the definition that $\mathbf{F}_r^m$ is a projector on $E^a(J^2)$. The set of *frequencies* of $\mathbf{F}_r^m$ is

$$\mathfrak{h}_r^m \;:=\; \{ \; (j,k) \in \mathbb{Z}^2 : \; |j| \le m^u \; , \; |k| \le m^v \; , \; u+v = r \; \} \quad .$$

Recall that

$$\mathfrak{h}_{m^r} \;:=\; \{ \; (j,k) \in \mathbb{Z}^2 : \; |j| \; , \; |k| \le m^r \; , \; |jk| \le m^r \; \}$$

which is the set of frequencies of the hyperbolic Fourier projector H_{m^r} . It is easily seen that the inclusion

$$\mathfrak{h}_r^m \;\subseteq\; \mathfrak{h}_{m^r}$$

holds. This implies the estimate

$$\| f - H_{m^r}(f) \| \;\le\; \| f - \mathbf{F}_r^m(f) \| \qquad (\; f \in L^2(J^2) \;) \; .$$

Proposition 1. *If* $f \in E^a(J^2)$ *and* $m \in \mathbb{N}$ *with* $m > 1$, *then*

$$(2) \qquad \|f - \mathbf{F}^m_r(f)\| = \mathcal{O}(r^{1/2}\cdot(m^r)^{-a+1/2}) \qquad (r \to \infty) .$$

Proof: The parametrically extended Fourier projectors satisfy

$$F'^{0}_{m} \le F'^{1}_{m} \le \cdots \le F'^{r}_{m} , \quad F''^{0}_{m} \le F''^{1}_{m} \le \cdots \le F''^{r}_{m} .$$

Therefore we can apply Proposition 2 of Section 1.4 and obtain

$$(3) \qquad (\mathbf{F}^m_r)^c$$

$$= (F'^{r}_{m})^c + (F'^{r-1}_{m})^c(F''^{0}_{m})^c + \cdots + (F'^{0}_{m})^c(F''^{r-1}_{m})^c + (F''^{r}_{m})^c$$

$$- \left((F'^{r}_{m})^c\cdot(F''^{0}_{m})^c + \cdots + (F'^{0}_{m})^c\cdot(F''^{r}_{m})^c \right)$$

which yields the *orthogonal decomposition* of $(\mathbf{F}^m_r)^c$:

$$(4) \qquad (\mathbf{F}^m_r)^c = (F'^{r}_{m})^c\cdot F''^{0}_{m} + \sum_{l=1}^{r} (F'^{r-l}_{m})^c\cdot(F''^{l}_{m} - F''^{l-1}_{m}) + (F''^{r}_{m})^c .$$

Then we can conclude

$$\|f - \mathbf{F}^m_r(f)\|^2$$

$$\le \|(F'^{r}_{m})^c\cdot F''^{0}_{m}(f)\|^2 + \sum_{l=1}^{r} \|(F'^{r-l}_{m})^c(F''^{l}_{m} - F''^{l-1}_{m})(f)\|^2 + \|(F''^{r}_{m})^c(f)\|^2$$

$$\le \|(F'^{r}_{m})^c(f)\|^2 + \sum_{l=1}^{r} \|(F'^{r-l}_{m})^c\cdot(F''^{l-1}_{m})^c(f)\|^2 + \|(F''^{r}_{m})^c(f)\|^2 .$$

Recall that

$$\|F'^{c}_{m}(f)\|^2 = \mathcal{O}(m^{-2a+1}) \quad (m \to \infty) \quad , \quad \|F'^{c}_{n}(f)\|^2 = \mathcal{O}(n^{-2a+1}) \quad (n \to \infty),$$

$$\|F'^{c}_{m}F''^{c}_{n}(f)\|^2 = \mathcal{O}(m^{-2a+1}\cdot n^{-2a+1}) \quad (m \to \infty \; , \; n \to \infty).$$

Then we obtain

$$\|f - \mathbf{F}^{m}_{r}(f)\|^2 = \mathcal{O}((r+2)\cdot(m^{r-1})^{-2a+1}) \quad (r \to \infty)$$

which implies

$$\|f - \mathbf{F}^{m}_{r}(f)\| = \mathcal{O}(r^{1/2}\cdot(m^{r})^{-a+1/2}) \quad (r \to \infty) .$$

This completes the proof of Proposition 1.

It follows from Proposition 1 of Section 7.7 that

$$\|f - H_{m^r}(f)\| = \mathcal{O}(\log(m^r)^{1/2}\cdot(m^{r})^{-a+1/2}) \quad (r \to \infty) .$$

Since $\log(m^r) = r\cdot\log(m)$, we get

(5) $$\|f - H_{m^r}(f)\| = \mathcal{O}(r^{1/2}\cdot(m^{r})^{-a+1/2}) \quad (r \to \infty) .$$

It follows from the relations (5) and (2) that the bivariate projectors H_{m^r} of hyperbolic Fourier approximation and $\mathbf{F}^{m}_{r}$ of pseudo-hyperbolic Fourier approximation possess the same order of approximationi $E^a(J^2)$. Recall that

$$\|f - H_{m^r}(f)\| \leq \|f - \mathbf{F}^{m}_{r}(f)\|$$

An application of Proposition 1 yields a proof of relation (5) via Boolean methods .

We consider now some special cases of Proposition 1.

Proposition 2. *If* $f \in \mathscr{C}^{1,1}(J^2)$, *then*

$$\|f - \mathbf{F}_r^m(f)\| = \mathcal{O}(r^{1/2} \cdot (m^r)^{-1/2}) \quad (r \to \infty) \tag{6}$$

Proof: It follows from Proposition 5 of Section 7.3 that $f \in E^1(J^2)$. An application of Proposition 1 yields relation (6) . This completes the proof of Proposition 2.

Proposition 3. *If* $f \in \mathscr{C}^{q+1,q+1}(J^2) \cap \mathscr{C}_{2\pi}^{q-1,q-1}(J^2)$, *then*

$$\|f - \mathbf{F}_r^m(f)\| = \mathcal{O}(r^{1/2} \cdot (m^r)^{-q-1/2}) \quad (r \to \infty) \quad . \tag{7}$$

Proof: It follows from Proposition 5 of Section 7.3 that $f \in E^{q+1}(J^2)$. An application of Proposition 1 yields relation (7). This completes the proof of Proposition 3.

We can derive error estimates in the uniform norm for $a > 1$.

Proposition 4. *If* $f \in E^a(J^2)$ *with* $a > 1$, *then*

$$\|f - \mathbf{F}_r^m(f)\|_\infty = \mathcal{O}(r \cdot (m^r)^{-a+1}) \quad (r \to \infty) \; . \tag{8}$$

Proof: Recall that

$$E^a(J^2) \subseteq A(J^2) \quad (a > 1) \quad .$$

Taking into account relation (4) we obtain

$$\|f - \mathbf{F}_r^m(f)\|_\infty$$

$$\leq \|(F'_m r)^c \cdot F''_m o(f)\|_\infty + \sum_{l=1}^{r} \|(F'_m r{-}1)^c \cdot (F''_m l - F''_m l{-}1)(f)\|_\infty + \|(F''_m r)^c (f)\|_\infty$$

$$\leq \sum_{|j|>m^r} \sum_{k=-\infty}^{\infty} |(f,e_{j,k})| + \sum_{l=1}^{r} \sum_{|j|>m^{r-l}} \sum_{|k|>m^{l-1}} |(f,e_{j,k})|$$

$$+ \sum_{j=-\infty}^{\infty} \sum_{|k|>m^r} |(f,e_{j,k})| \quad .$$

Recall that

$$\sum_{|j|>m} \sum_{|k|>n} |(f,e_{j,k})| = \mathcal{O}(m^{-a+1} \cdot n^{-a+1}) \quad (m \to \infty,\ n \to \infty) \ .$$

Then we can conclude

$$\|f - F_r^m(f)\|_\infty = \mathcal{O}(r \cdot (m^r)^{-a+1}) \quad (r \to \infty)$$

which completes the proof of Proposition 4 .

Proposition 5. *If* $f \in \mathcal{C}^{q+1,q+1}(J^2) \cap \mathcal{C}_{2\pi}^{q-1,q-1}(J^2)$, *then*

(9) $$\|f - F_r^m(f)\|_\infty = \mathcal{O}(r \cdot (m^r)^{-q}) \quad (r \to \infty)$$

Proof: It follows from Proposition 5 of Section 7.3 that $f \in E^{q+1}(J^2)$. An application of Proposition 4 yields relation (9). This completes the proof of Proposition 5.

Notes, Section 7

Fourier blending as a bivariate approximation tool was first mentioned in GORDON [27]. The Korobov spaces are of basic importance in multidimensional numerical quadrature (see HUA - WANG [33] and SLOAN [43]). Our representation of Boolean Fourier approximation follows BASZENSKI - DELVOS [4] (see also BASZENSKI [1] and DELVOS [14]) . We also refer to DIN'ZUNG [26].

8 Boolean cubature

8.1 Splines and the trapezoidal rule

We adopt the notations of Section 6. Let $N \in \mathbb{N}$ and S_N be the projector of periodic spline interpolation of degree 2p-1 ($p \in \mathbb{N}$)with respect to the uniform mesh

$$x_j = j \cdot h_N \quad (j \in \mathbb{Z} \ , \ h_N = 2\pi/N) .$$

We have the Lagrange representation

$$(1) \qquad S_N(f)(x) = \sum_{j=0}^{N-1} f(jh_N) \cdot L_N(x - jh_N)$$

where L_N is the fundamental periodic spline function satisfying

$$L_N(j \cdot h_N) = \delta_{0,j} \quad (0 \le j < N) .$$

Proposition 1. *If* $f \in \mathscr{C}_{2\pi}$, *then*

$$(2) \qquad T_N(f) = \frac{1}{2\pi} \int_0^{2\pi} S_N(f)(x)\, dx = \frac{1}{N} \sum_{j=0}^{N-1} f(jh_N) .$$

Thus, the trapezoidal (or rectangle) rule for periodic functions is a spline interpolatory quadrature formula.

Proof: Taking into account relation (1) we obtain

$$\frac{1}{2\pi} \int_0^{2\pi} S_N(f)(x)\, dx = (S_N(f), e_0) = \sum_{j=0}^{N-1} f(jh_N) \cdot a_j .$$

with

$$a_j = \frac{1}{2\pi}\int_0^{2\pi} L_N(x-jh_N)\,dx = \frac{1}{2\pi}\int_0^{2\pi} L_N(x)\,dx = a_0 \quad (0 \le j < N).$$

Since $S_N(e_0) = e_0$, we obtain

$$a_j = \frac{1}{N} \quad (0 \le j < N) .$$

which completes the proof of Proposition 1.

Proposition 2. *If* $f \in \mathcal{C}_{2\pi}^{2p}$ *with* $p \in \mathbb{N}$ *, then*

(3) $$T_N(f) - \frac{1}{2\pi}\int_0^{2\pi} f(x)\,dx = \mathcal{O}(N^{-2p}) \quad (N \to \infty) .$$

Proof: We have

$$T_N(f) - \frac{1}{2\pi}\int_0^{2\pi} f(x)\,dx = (S_N(f)-f,e_0) .$$

An application of Proposition 2 in Section 6.1 yields (3) which completes the proof of Proposition 2.

Recall that $S_N(f) \in \mathcal{C}_{2\pi}^{2p-2} \cap \mathcal{C}^{2p}(J)$. It follows from Proposition 8 of Section 7.1 that $S_N(f) \in E^{2p}(J)$. The following result relaxes the smoothness condition of Proposition 2.

Proposition 3. *If* $f \in E^a(J)$ *with* $a > 1$ *, then*

(4) $$T_N(f) - \frac{1}{2\pi}\int_0^{2\pi} f(x)\,dx = \mathcal{O}(N^{-a}) \quad (N \to \infty) .$$

Proof: Note first that $E^a(J) \subseteq A(J)$. This implies

$$(5) \qquad -(S_N{}^c(f),e_0) = T_N(f) - \frac{1}{2\pi}\int_0^{2\pi} f(x)\,dx = \sum_{k\neq 0} (f,e_{kN}) \ .$$

It follows from $f \in E^a(J)$ that

$$(f,e_k) = \mathcal{O}(|k|^{-a}) \quad (|k| \to \infty) \ .$$

Relation (4) is now an easy consequence of of (5) and $a > 1$ which completes the proof of Proposition 3 .

8.2 Bivariate splines and the tensor product trapezoidal rule

Let m and N be positive integers. The parametrically extended spline projectors S'_M and S''_N are given by

$$S'_M(f)(x,y) = \sum_{j=0}^{M-1} f(jh_M,y)L_M(x-jh_M), \qquad S''_N(f)(x,y) = \sum_{k=0}^{N-1} f(x,kh_N)L_N(y-kh_N).$$

The projector of bivariate tensor product spline interpolation possesses the representation

$$S'_M S''_N(f)(x,y) = \sum_{j=0}^{M-1}\sum_{k=0}^{N-1} f(jh_M,kh_N)\cdot L_M(x-jh_M)\cdot L_N(y-kh_N) \quad .$$

As in the univariate case the bivariate tensor product trapezoidal rule is obtained by integrating the bivariate tensor product spline interpolant $S'_M S''_N(f)$:

$$T_{M,N}(f) = \left(\frac{1}{2\pi}\right)^2 \int_0^{2\pi}\int_0^{2\pi} S'_M S''_N(f)(x,y)\, dx\, dy \quad .$$

We have

$$(1) \qquad T_{M,N}(f) = \frac{1}{M\cdot N}\sum_{j=0}^{M-1}\sum_{k=0}^{N-1} f(jh_M,kh_N) \quad .$$

Proposition 1. *If* $f \in \mathscr{C}_{2\pi}^{2p,2p}$ *with* $p \in \mathbb{N}$, *then*

$$(2) \quad T_{M,N}(f) - \left(\frac{1}{2\pi}\right)^2 \int_0^{2\pi}\int_0^{2\pi} S'_M S''_N(f)(x,y)\, dx\, dy = \mathcal{O}(M^{-2p}+N^{-2p}) \quad (M,N \to \infty).$$

Proof: As in the univariate case we have

$$(3)\qquad T_{M,N}(f) - \left(\frac{1}{2\pi}\right)^2 \int_0^{2\pi}\int_0^{2\pi} S'_M S''_N(f)(x,y)\,dx\,dy = (S'_M S''_N(f)-f,e_{0,0}) .$$

Moreover, the relation

$$(4)\qquad (S'_M S''_N(f)-f,e_{0,0})$$

$$= -(S'^c_M(f)-f,e_{0,0}) - (S''^c_N(f)-f,e_{0,0}) + (S'^c_M S''^c_N(f)-f,e_{0,0})$$

holds. It can be shown as in Section 6.2 that

$$\|f-S'_M(f)\|_\infty = \mathcal{O}(M^{-2p}) , \quad \|f-S''_N(f)\|_\infty = \mathcal{O}(N^{-2p}) \quad (M,N \to \infty),$$

$$\|f-S'_M \oplus S''_N f)\|_\infty = \mathcal{O}(M^{-2p}\cdot N^{-2p}) \quad (M,N \to \infty)$$

for $f \in \mathcal{C}^{2p,2p}_{2\pi}$. Applying these relations to (4) we obtain (2) which completes the proof of Proposition 1.

It is easily seen that $S'_M S''_N(f) \in \mathcal{C}^{2p-2,2p-2}_{2\pi} \cap \mathcal{C}^{2p,2p}(J^2)$. It follows from Proposition 8 of Section 7.3 that

$$S'_M S''_N(f) \in E^{2p}(J^2) .$$

The following result relaxes the smoothness condition of Proposition 1.

Proposition 2. *If* $f \in E^a(J^2)$ *with* $a > 1$, *then*

$$(5)\qquad T_{M,N}(f) - \left(\frac{1}{2\pi}\right)^2 \int_0^{2\pi}\int_0^{2\pi} S'_M S''_N(f)(x,y)\,dx\,dy = \mathcal{O}(M^{-a}+N^{-a}) \quad (M,N \to \infty).$$

Proof: Note first that $A(J^2) \subseteq E^a(J^2)$. This implies

$$-(S'^{c}_{M}(f)-f,e_{0,0}) = \sum_{k\neq 0}(f,e_{kM,0}), \qquad -(S''^{c}_{N}(f)-f,e_{0,0}) = \sum_{l\neq 0}(f,e_{0,lN}),$$

$$(S'^{c}_{M}S''^{c}_{N}(f)-f,e_{0,0}) = \sum_{k\neq 0}\sum_{l\neq 0}(f,e_{kM,lN}) \quad .$$

It follows from $f \in E^{a}(J^{2})$ that

$$(f,e_{k,l}) = \mathcal{O}(|k|^{-a}\cdot|l|^{-a}) \quad (\ |k|,|l| \to \infty\ ) \quad .$$

Thus we can conclude

(6) $$-(S'^{c}_{M}(f)-f,e_{0,0}) = \mathcal{O}(M^{-a}) \quad (\ M \to \infty\ )\ ,$$

(7) $$-(S''^{c}_{N}(f)-f,e_{0,0}) = \mathcal{O}(N^{-a}) \quad (\ N \to \infty\ )\ ,$$

(8) $$(S'^{c}_{M}S''^{c}_{N}(f)-f,e_{0,0}) = \mathcal{O}(M^{-a}\cdot N^{-a}) \quad (\ M,N \to \infty\ )\ .$$

Relation (5) is now an direct consequence of (6), (7), (8), and (4). This completes the proof of Proposition 2.

8.3 r-th order blending cubature

The projector of bivariate r-th order spline blending interpolation is given by

$$(1)\qquad \mathbf{B}_r^2 = \sum_{m=1}^{r} S'_{2^m} S''_{2^{r-m+1}} - \sum_{m=1}^{r-1} S'_{2^m} S''_{2^{r-m}} .$$

We define the *r-th order blending cubature formula* as the integral of $\mathbf{B}_r^2(f)$:

$$\mathbf{T}_r^2(f) = \left(\frac{1}{2\pi}\right)^2 \int_0^{2\pi}\int_0^{2\pi} \mathbf{B}_r^2(f)(x,y)\,dx\,dy .$$

It follows from relation (1) that

$$(2)\qquad \mathbf{T}_r^2(f) = \mathbf{S}_r^2(f) - \mathbf{S}_{r-1}^2(f)$$

with

$$(3)\qquad \mathbf{S}_r^2(f) = \sum_{m=1}^{r} T_{2^m,2^{r-m+1}}(f) .$$

The number of point evaluations for computing $\mathbf{T}_r^2(f)$ is mainly determined by the number of points occuring in (3) which is given by

$$n_r = (r+1)\cdot 2^r$$

(see Section 6.3).

Proposition 1. *If* $f \in \mathcal{C}_{2\pi}^{2p,2p}$ *with* $p \in \mathbb{N}$ *, then*

$$(4)\qquad \mathbf{T}_r^2(f) - \left(\frac{1}{2\pi}\right)^2 \int_0^{2\pi}\int_0^{2\pi} f(x,y)\,dx\,dy = \mathcal{O}(r\cdot(2^r)^{-2p}) \quad (r \to \infty) .$$

Proof: Since

$$T_r^2(f) - \left(\frac{1}{2\pi}\right)^2 \int_0^{2\pi}\int_0^{2\pi} B_r^2(f)(x,y)\,dx\,dy = (T_r^2(f)-f,e_{0,0}) ,$$

an application of Proposition 3 in Section 6.3 yields (4). This completes the the proof of Proposition 1.

Again we will relax the smoothness assumption of Proposition 1.

Proposition 2. *If* $f \in E^a(J^2)$ *with* $a > 1$, *then*

$$(5) \qquad T_r^2(f) - \left(\frac{1}{2\pi}\right)^2 \int_0^{2\pi}\int_0^{2\pi} f(x,y)\,dx\,dy = \mathcal{O}(r\cdot(2^r)^{-a}) \quad (r \to \infty) .$$

Proof: We have in view of relation (1)

$$(6) \quad T_r^2(f) - (f,e_{0,0})$$

$$= \sum_{m=1}^{r}\left(T_{2^m,2^{r-m+1}}(f) - (f,e_{0,0})\right) - \sum_{m=1}^{r-1}\left(T_{2^m,2^{r-m}}(f) - (f,e_{0,0})\right) .$$

We have shown in Section 8.2 that

$$(7) \qquad T_{2^u,2^v}(f) - (f,e_{0,0})$$

$$= -(S'^c_{2^u}(f),e_{0,0}) - (S''^c_{2^v}(f),e_{0,0}) + (S'^c_{2^u}S''^c_{2^v}(f),e_{0,0}).$$

Taking into account (6) and (7) we obtain

$$T_r^2(f) - (f,e_{0,0})$$

$$= \sum_{m=1}^{r} (S'_{2^m}{}^c S''_{2^{r-m+1}}{}^c(f), e_{0,0})$$

$$- \sum_{m=1}^{r} (S'_{2^m}{}^c(f), e_{0,0}) - \sum_{m=1}^{r} (S''_{2^{r-m+1}}{}^c(f), e_{0,0})$$

$$- \sum_{m=1}^{r-1} (S'_{2^m}{}^c S''_{2^{r-m}}{}^c(f), e_{0,0})$$

$$+ \sum_{m=1}^{r-1} (S'_{2^m}{}^c(f), e_{0,0}) + \sum_{m=1}^{r-1} (S''_{2^{r-m}}{}^c(f), e_{0,0}) ,$$

i. e., we have

$$(8) \quad \mathbf{T}_r^2(f) - (f, e_{0,0})$$

$$= \sum_{m=1}^{r} (S'_{2^m}{}^c S''_{2^{r-m+1}}{}^c(f), e_{0,0})$$

$$- \sum_{m=1}^{r-1} (S'_{2^m}{}^c S''_{2^{r-m}}{}^c(f), e_{0,0}) - (S'_{2^r}{}^c(f), e_{0,0}) - (S''_{2^r}{}^c(f), e_{0,0}) .$$

An application of the relations (6), (7), (8) of Section 8.2 to relation (8) yields relation (5) . This completes the proof of Proposition 2.

We will apply the bivariate r-th order blending cubature formula to approximate the double integral

$$\int_0^1 \int_0^1 \frac{x + y}{1 + x \cdot y} \, dx \, dy = 2(\log(4) - 1).$$

Using the function

$$f(x,y) = 2\pi(x + y)/(4\pi^2 + x \cdot y)$$

we obtain

$$\left(\frac{1}{2\pi}\right)^2 \int_0^{2\pi} \int_0^{2\pi} f(x,y)\, dx\, dy \quad = \quad 2(\log(4) - 1).$$

Since $f \in E^1(J^2)$, we consider the function

$$g(x,y) \quad = \quad \frac{1}{4}\Big(f(x,y) + f(x,2\pi-y) + f(2\pi-x,y) + f(2\pi-x,2\pi-y)\Big)$$

which is element of $E^2(J^2)$ and satisfies

$$\left(\frac{1}{2\pi}\right)^2 \int_0^{2\pi} \int_0^{2\pi} g(x,y)\, dx\, dy \quad = \quad 2(\log(4) - 1).$$

Thus, Proposition 2 is applicable with $a = 2$.

Table 1 shows the errors for the r-th order blending cubature while Table 2 contains the same information for the tensor product trapezoidal rule.

r	no of terms	$\mathbf{T}_r^2(g) - (g,e_{0,0})$
1	4	.01009
2	12	.00365
3	32	.00120
4	80	.00037
5	192	.00011
6	448	.00003

Table 1

r	no of terms	$\mathbf{T}_{2^r,2^r}(g) - (g,e_{0,0})$
1	4	.01009
2	16	.00282
3	64	.00072
4	256	.00018
5	1024	.00005
6	4096	.00001

Table 2

Notes, Section 8

We refer to HUA - WANG [33] and SLOAN [43] for multivariate lattices rules and corresponding error estimates in Korobov spaces. It follows from relation (5) that r-th order blending cubature is comparable with bivariate good lattices rules.

References

1 G.BASZENSKI: "Zur Konvergenzbeschleunigung von Orthogonal-Doppelreihen". Dissertation, Siegen, 1983.

2 G.BASZENSKI: N-th order polynomial spline blending. *In* "Multivariate Approximation Theory III" (W. Schempp, K. Zeller, Eds.), pp. 35-46. ISNM **75**, Birkhäuser, Basel, 1985.

3 G. BASZENSKI and F.-J. DELVOS: Acceerating the rate of convergence of bivariate Fourier expansions. *In* "Approximation Theory IV" (C. K. Chui, L. L. Schumaker, J. D. Ward, Eds.), pp. 335-340. Academic Press, New York, 1983.

4 G. BASZENSKI and F.-J. DELVOS: Boolean methods in Fourier approximation. *In* "Topics in Multivariate Approximation" (C. K. Chui, L. L. Schumaker, F. Utreras, Eds.), pp. 1-12, Academic Press, New York, 1987.

5 G.BASZENSKI and F.-J. DELVOS: Boolean algebra and multivariate interpolation. *In* "Approximation and function spaces", Banach Center Publications **22** , Polish Scientific Publishers, Warsaw, 1988. (to appear).

6 G.BASZENSKI, F.-J. DELVOS, and H. POSDORF: Boolean methods in bivariate reduced Hermite interpolation. *In* "Multivariate Approximation Theory " (W. Schempp, K. Zeller, Eds.), pp. 11-30. ISNM **51**, Birkhäuser, Basel, 1979 .

7 G.BASZENSKI,F.-J. DELVOS, and K. HACKENBERG: Remarks on reduced Hermite interpolation. *In* "Multivariate Approximation Theory III" (W. Schempp, K. Zeller, Eds.), pp. 47-58. ISNM **61**, Birkhäuser, Basel, 1982.

8 O. BIERMANN: Über näherungsweise Cubaturen. Monatshefte für Mathematik und Physik **14**(1903), 211-225.

9 G. BIRKHOFF: The algebra of multivariate interpolation. *In* "Constructive approaches to mathematical models" (Eds. : C. W. Coffman and G. Fix), pp. 347-363, Academic Press (1979), New York **1979** .

10 G.BIRKHOFF and W. J. GORDON: The draftman's and related equations. Journal of Approximation Theory **1**(1968), 199-208.

11 F.-J. DELVOS: On surface interpolation. Journal of Approximation Theory **15**(1975), 209-213.

12 F.-J. DELVOS: d-variate Boolean interpolation. Journal of Approximation Theory **34**(1982), 99-114.

13 F.-J. DELVOS : Intermediate blending interpolation. *In* "Multivariate Approximation Theory III" (W. Schempp, K. Zeller, Eds.), pp. 35-46. ISNM **75**, Birkhäuser, Basel, 1985.

14 F.-J. DELVOS: Fourier series and right invertible operators. Mathematische Nachrichten **122**(1985), 99-108.

15 F.-J. DELVOS: Splines and pseudo-inverses. RAIRO Analyse Numérique **12**(1978), 313-324.

16 F.-J. DELVOS: Limits of abstract splines. *In* "Inverse and Ill-Posed Problems" (H. W. Engl and C. W. Groetsch, Eds.), Academic Press, New York, 1987.

17 F.-J. DELVOS: On the convergence of periodic splines of arbitrary degree. Journal of Approximation Theory **53**(1988), 221-230.

18 F.-J. DELVOS and H. POSDORF: On optimal tensor product approximation. Journal of Approximation Theory **18**(1976), 99-107.

19 F.-J. DELVOS and H. POSDORF: N-th order blending. *In* "Constructive Theory of Functions of Several Variables" (W. Schempp, K. Zeller, Eds.), pp. 53-64. Lecture Notes in Mathematics **571**, Springer, Heidelberg, 1977.

20 F.-J. DELVOS and H. POSDORF: A representation formula for reduced Hermite interpolation. *In* "Numerische Methoden der Approximationstheorie, Bd. 4" (L. Collatz, G. Meinardus, H. Werner, Eds.), pp. 124-137. ISNM **42**, Birkhäuser, Basel, 1978.

21 F.-J. DELVOS and H. POSDORF: Boolesche zweidimensionale Lagrange - Interpolation. *Computing* **22** (1979), 311-323.

22 F.-J. DELVOS and H. POSDORF: Generalized Biermann interpolation. *Resultate Math.* **5** (1982), 6-18.

23 F.-J. DELVOS and W. SCHEMPP: On precision sets of interpolation projectors. *In* "Multivariate Approximation Theory III" (W. Schempp, K. Zeller, Eds.), pp. 107 - 124, ISNM **61**, Birkhäuser, Basel, 1982 .

24 F.-J. DELVOS and W. SCHEMPP: The method of parametric extension applied to right invertible operators. *Numer. Funct. Anal. Optim* **6** (1983), 135-148.

25 F.-J. DELVOS and W. SCHEMPP: Interpolation projectors and closed ideals. *In* "Approximation and function spaces", Banach Center Publications **22** , Polish Scientific Publishers, Warsaw, 1988. (to appear).

26 DIN'ZUNG: Approximation by trigonometric polynomials of functions of several variables on the Torus. Math. USSR Sbornik 59(1988), 247-267.

27 W.J. GORDON: Spline-blended surface interpolation through curve networks. *J. Math. Mech.*, Vol.**18**, No. 10 (1969), 931-952.

28 W.J. GORDON: Distributive lattices and approximation of multivariate functions. *In* "Proc. Symp. Approximation with Special Emphasis on Spline Functions (Madison, Wisc. 1969)" (I.J. Schoenberg, Ed.), pp. 223-277.

29 W.J. GORDON: Blending function methods of bivariate and multivariate interpolation and approximation. *SIAN, J. Numer. Anal.*, Vol. **8**, No. 1 (1971), 158-177.

30 W. J. GORDON and E. W. CHENEY: Bivariate and multivariate interpolation with noncommutative projectors. *In* "Linear Spaces and Approximation" (P. L. Butzer and B. Sz. Nagy, Eds.), pp. 381-387, ISNM **40**, Birkhäuser, Basel, 1977.

31 W.J. GORDON, C.A. HALL: Transfinite element methods: blending function interpolation over arbitrary curved element domains. *Numer. Math* **21** (1973), 109-129.

32 W. J. GORDON and J. A. WIXOM : Pseudo-harmonic interpolation on convex domains. SIAM J. Numer. Anal. 11.(1974), 909-933.

33 HUA LOO KENG and WANG YUAN: "Applications of Number Theory to Numerical Analysis" . Springer-Verlag, Heidelberg 1981 .

34 N.P. KORNEĬCUK: Exact error bounds of approximation by interpolating splines in L-metric on the classes W_p^r $(1 \le p < \infty)$ of periodic functions. *Anal. Math.* **3** (1977), 109-117.

35 P. LANCASTER: Interpolation in a rectangle and finite elements of high degree. *J. Inst. Math. Appl.* **18** (1976), 67-77.

36 P. LANCASTER and D.S. WATKINS: Some families of finite elements. *J. Inst. Math. Appl.* **19** (1977), 385-397.

37 W.A. LIGHT and E.W. CHENEY: "Approximation theory in tensor product spaces". Springer, Heidelberg, 1985.

38 F. MELKES: Reduced piecewise bivariate Hermite interpolations. *Numer. Math.* **19** (1972), 326-340.

39 L. NACHBIN: "Elements of Approximation Theory" , R. E. Krieger Publishing Co., Huntington, 1976

40 H. POSDORF: "Boolesche Methoden bei zweidimensionaler Interpolation". Dissertation, Siegen, 1978.

41 D. PRZEWORSKA-ROLEWICZ: "Algebraic analysis". D. Reidel Publishing Company, Dordrecht, 1988 .

42 Z. SEMADENI: "Banach spaces of continuous functions, vol.", Monogr. Math. **55**, Polosh Scientific Publishers, Warsaw 1971 .

43 I. K. SLOAN: Lattice methods for multiple integration. Journal of Computional and Applied Mathematics **12-13**(1985), 131-143 .

44 D.D. STANCU, The remainder of certain linear approximation formulas in two variables. *SIAM J. Numer. Anal.*, Ser. **B 1** (1964), 137-163.

45 M. TASCHE: A unified approach to interpolation methods. Journal of Integral Equations **4** (1982), 55-75 .

46 D.S. WATKINS: A conforming rectangular plate element. *In* "The Mathematics of Finite Elements and Applications II, MAFELAP 1975" (J.R. Whiteman, Ed.), pp. 77-83. Academic Press, New York, 1976.

47 D.S. WATKINS: On the construction of conforming rectangular plate elements. *Internat. J. Numer. Methods Engrg.* **10** (1976), 925-933.

48 O.C. ZIENKIEWICZ: "The finite element method in engineering science". McGraw-Hill, New York, 1971.